Eugene Roventa and Tiberiu Spircu

Management of Knowledge Imperfection in Building Intelligent Systems

Studies in Fuzziness and Soft Computing, Volume 227

Editor-in-Chief

Prof. Janusz Kacprzyk
Systems Research Institute
Polish Academy of Sciences
ul. Newelska 6
01-447 Warsaw
Poland
E-mail: kacprzyk@ibspan.waw.pl

Further volumes of this series can be found on our homepage: springer.com

Eugene Roventa and
Tiberiu Spircu

Management of Knowledge Imperfection in Building Intelligent Systems

 Springer

Authors

Prof. Eugene Roventa
Glendon Computer Science and IT Department
Glendon College
York University
2275 Bayview Avenue
Toronto, Ontario M4N 3M6
Canada
E-mail: roventa@yorku.ca

Prof. Tiberiu Spircu
Dept. of Medical Informatics
"Carol Davila" University of Medicine and Pharmacy
8 Eroii Sanitari Blvd.
Bucharest
Romania

ISBN 978-3-540-77462-4   e-ISBN 978-3-540-77463-1

DOI 10.1007/978-3-540-77463-1

Studies in Fuzziness and Soft Computing   ISSN 1434-9922

Library of Congress Control Number: 2008933226

© 2009 Springer-Verlag Berlin Heidelberg

Typeset & Cover Design: Scientific Publishing Services Pvt. Ltd., Chennai, India.

Printed in acid-free paper

9 8 7 6 5 4 3 2 1

springer.com

Preface

There are many good AI books. Usually they consecrate at most one or two chapters to the imprecision knowledge processing. To our knowledge this is among the few books to be entirely dedicated to the treatment of knowledge imperfection when building intelligent systems. We consider that an entire book should be focused on this important aspect of knowledge processing. The expected audience for this book includes undergraduate students in computer science, IT&C, mathematics, business, medicine, etc., graduates, specialists and researchers in these fields. The subjects treated in the book include expert systems, knowledge representation, reasoning under knowledge Imperfection (Probability Theory, Possibility Theory, Belief Theory, and Approximate Reasoning). Most of the examples discussed in details throughout the book are from the medical domain. Each chapter ends with a set of carefully pedagogically chosen exercises, which complete solution provided. Their understanding will trigger the comprehension of the theoretical notions, concepts and results.

Chapter 1 is dedicated to the review of expert systems. Hence are briefly discussed production rules, structure of ES, reasoning in an ES, and conflict resolution. Chapter 2 treats knowledge representation. That includes the study of the differences between data, information and knowledge, logical systems with focus on predicate calculus, inference rules in classical logic, semantic nets and frames. Chapter 3 starts with taxonomy of imperfection, usual and precise meaning, and continues with experiments and events, formal definition of events, defining probabilities, Bayes Theorem, random variables and distributions, expectation and variance, and examples. Chapter 4 deals with statistical inference. Chapter 5 is dedicated to Bayesian networks. It treats uncertain production rules, belief networks with examples, and different software to implement them. Chapter 6 tackles the certainty factors theory. Belief theory is discussed in Chapter 7. Thus, belief approach, agreement measures, Dempster-Shaffer theory, combining beliefs topics, and transferable belief model are highlighted. Chapter 8 reviews some aspects of possibility theory. Approximate reasoning, with the sub-topics fuzzy sets, fuzzy logic, hedges, and defuzzyfication are the key words for the Chapter 9.

The next and the last chapter, Chapter 10, is dedicated to a short review of different usable notions in uncertainty management, as well as to the presentation of the Computational Theory of Perception of Zadeh.

Obviously, the authors take full responsibility for all opinions and mistakes, either hidden or not through the book. (Such as, for example, to leave out any reference to neural nets.) They will appreciate, however, if the readers will send back to them their comments and suggestions.

Features of the book:

a) Comprehensive comparative approach to deal with most of the techniques of management of knowledge imperfection
b) Breakthrough fuzzy techniques approach for handling real word imprecision
c) Numerous examples throughout the book in the medical domain
d) Each chapter is followed by a set of detailed solved exercises.

July 2008   The Authors

Notations

2^{Ω}	power set of Ω
$\mathcal{A}$	algebra
	arrow set of a directed acyclic graph
α	significance level, precision degree
$\mathcal{B}$	the Borel sigma-algebra
$b(n, \pi)$	binomial distribution, parameters n and π
$Be(\pi)$	Bernoulli distribution, parameter π
$\mathrm{Bel}(A)$	belief in subset A of worlds
$\mathrm{bel}_{\Omega}(\bullet \mid X)$	conditional belief measures
cf	certainty factor
$Cov(X, Y)$	covariance of a pair of random variables X and Y
D	descriptor
$Dom(V)$	domain of variable V
E	event
$\overline{E}$	complement of event E
$E(X)$, $E(\phi)$	expectation of random variable X, of function ϕ
EA	expected agreement
$F : \mathbf{R} \rightarrow [0, 1]$	distribution function
$g : \mathcal{P} \rightarrow [0, 1]$	uncertainty measure
H	hedge
κ	kappa coefficient
$m(\omega)$	measure of "world" ω
$m(A)$	measure of subset A of worlds
MB	measure of belief
MD	measure of disbelief
$\mu_F : \Omega \rightarrow [0, 1]$	membership function of fuzzy set F over Ω
$\mathbf{N}$	set of natural numbers
$\mathcal{N}$	node set of a directed acyclic graph

$N(\mu,\sigma^2)$	normal distribution, parameters μ and σ^2
$N:\mathcal{P}\rightarrow[0,1]$	necessity measure
$N_{v\bullet}$, $N_{\bullet w}$	row total, column total in contingency tables
$N_{\bullet\bullet}$	total number of cases in a contingency table
$O(e)$	odds in favor of proposition e
OA	observed agreement
$P(E)$	probability of event E
$P(f)$	probability of formula f
$P_L(E)$	logical probability of event E
$p(A\mid B)$	probability of event A conditioned by event B
$p(h\mid e)$	probability of formula h conditioned by formula e
$Pl(A)$	plausibility of subset A of worlds
$Po(\lambda)$	Poisson distribution, parameter λ
$\Pi:\mathcal{P}\rightarrow[0,1]$	possibility measure
$\pi()$	predicate
$\pi:\Omega\rightarrow[0,1]$	possibility distribution
$\pi(q\mid p)$	conditional possibility of proposition q given p
$\mathbf{R}$	set of real numbers
$\mathbf{R}^m$	m-dimensional real space
$\rho(X,Y)$	correlation coefficient of two random variables
S	sigma-algebra
s^2	sample variance
$\Sigma\#(F)$	sigma-count of fuzzy set F
$t(\nu)$	Student (t) distribution, parameter ν
$Var(X)$	variance of random variable X
$x_{(i)}$	ordered component of a data sample
X^2	Pearson statistic
χ_A	characteristic function of subset A
$\chi^2(\nu)$	chi-square distribution, parameter ν
ω	world
Ω	universe
$\mid X\mid$	cardinal of set X
$\wedge$	logical "and"
$\vee$	logical "or"
$\Rightarrow$	logical "if-then" (implication)
$\Rightarrow_K$	Kleene implication

$\Rightarrow_H$	Heyting implication
$\Leftrightarrow$	logical equivalence
$\neg$	logical "not"
$\forall$	quantifier "any"
$\exists$	quantifier "exists"
$\mapsto$	"infer"
#	number
$\varnothing$	impossible event
$\cup$	disjunction of events, union of subsets, union of fuzzy sets
$\cap$	conjunction of events, intersection of subsets, intersection of fuzzy sets
$\displaystyle\bigcup_{n=1}^{\infty}$	countable union of subsets
$\displaystyle\bigcap_{n=1}^{\infty}$	countable intersection of subsets
$\times$	Cartesian product of fuzzy sets
$\otimes$	tensor product, multiplication of fuzzy numbers
$\oplus$	composition of beliefs, addition of fuzzy numbers
$\circ$	composition of functions, composition of (crisp, fuzzy) relations

Contents

1 "Classical" Expert Systems

1.1 Production Rules

Getting knowledge means both accumulating knowledge and understanding a subject or a domain (of knowledge).

The (human) experts are people who possess enough knowledge in a certain field, who acquired in the past a practical experience in that field and can give now pertinent advices in it. The experts are able to give definitions and/or concepts and also to express their knowledge in terms of rules.

Definitions are of the form:

$$term \quad IF \quad condition$$

The conditions can be composed of several elements connected by AND or OR. As an example, some experts may define the term "pneumonia" by the following condition:

temperature > 38°C   AND   white blood count > 10000 cells/mm^3
                     AND   chest X-ray = lobar infiltrate
                     AND   symptom = stitch.

Production rules are of the form:

IF	*antecedent*	THEN	*consequent*	
	premise		*conclusion*	(in logic)
	condition		*action*	(in practice)

The antecedent can be "multiple", i.e. composed of several elements connected by AND or OR. The consequent also can be "multiple". These components will be named **facts**.

The structure of a fact is very simple: it contains a variable – which represents a characteristic of an object, a possible value of this variable, and an operator that connects the variable and the value.

All this can be expressed either as a predicate

operator(variable, value)

or, in other form

variable_operator_value.

E. Roventa and T. Spircu: Management of Knowledge Imperfection, STUDFUZZ 227, pp. 1–12.
springerlink.com   © Springer-Verlag Berlin Heidelberg 2009

For example, the objects are Canadian citizens, which have as characteristics the height, the eye color, etc. Considering the height, it is possible to represent it by several variables, such as:

- The height expressed in *centimeters*. Here the centimeter was chosen as unit, and the values are obtained by measurements. Therefore we will obtain real numbers as values. Usually these numbers will be approximated by natural numbers.
- The height expressed in *feet* and *inches*. It is practically the same situation as above, but this time the values are presented in pairs (of natural numbers).
- The height expressed linguistically. When we do not have the time or the conditions to precisely measure a person, we can appreciate (subjectively) his height in terms such as "very tall", "tall", "medium" and so on. Here the values are fuzzy terms.

If we consider the eye color, then it seems that the variable representing this characteristic has to have linguistic values, the names of colors. Nevertheless, it is worth noting that numbers could represent colors (in optics just one – the wave-length, in imagistic three – the red, green and blue components).

Among the linguistic variables there is a group that is treated by specific means. These are the Boolean variables, which can take only two values: true res. false.

Now, the operators connecting the variables to their values are of mathematical type

$$=, <, \leq, \ldots$$

or of linguistic type

is, has, ...

A production rule can represent many things: simple relations, recommendations to follow, orders to be executed, research strategies, heuristics obtained from previous experience. Here are typical examples:

• relation	IF	prednisone is taken at a level of 30 mg/day
	THEN	cholesterol increases by 5 to 13 mg/l
• recommendation	IF	the number of patients affected by a disease is very small compared to the number of patients in the study
	AND	the disease is present throughout these records
	THEN	the advice is "eliminate these records from the study"
• directive [Negnevitsky 2000]	IF	the car is dead
	AND	the fuel tank is empty
	AND	we need to use the car
	THEN	refuel the car

	IF	heart rate is between 100 and 150 beats/minute
• strategy	THEN	significant tachycardia
	AND	heart rate was checked
	IF	heart rate was checked
	AND	PR interval is larger than 0.2 seconds
	THEN	1st degree AV block
	AND	PR interval checked
• heuristic	IF	the patient coughs
	AND	whizzing is present
	AND	dispnæa is present
	THEN	it is almost sure that 'asthma' is the 'disease of patient'

1.2 Expert Systems

Expert systems are computer programs for solving complex problems at the level of human expert in a narrow problem area.

Since a human expert continuously adapts his knowledge, it is clear that an expert system will be accepted only if:

− its knowledge base can be easily updated and/or corrected, and
− its conclusions can be justified.

An expert system is capable to function even its knowledge base is quite incomplete. Usually, the learning process during which the knowledge base is updated at an acceptable level is long and costly. The knowledge base has to contain thousand of rules to be fully operational.

Before the personal computer era (i.e. before 1980), the general opinion was that for developing an expert system it is necessary to form a large team composed by:

1) Project manager,
2) Domain expert(s),
3) Knowledge engineer(s),
4) Programmers.

The leader was the project manager, who organized the work, and coordinated the others members of the team.

The domain expert had the greatest expertise (to be captured in the expert system) − in a certain domain − and the ability to communicate his knowledge.

The programmers had to transform the acquired knowledge in computer programs (written in Prolog, LISP, or other specific programming languages).

The knowledge engineer was a person capable of designing, building, testing, and integrating an expert system in the workplace. He was a liaison between the domain expert and the programmers. Finally,

5) The end-user is any person who uses the expert system once it was developed.

Today this team is reduced since we can use expert system shells. This can eliminate the need for programmers and also might reduce the role of knowledge engineers.

An expert system shell is an expert system with the knowledge removed. Therefore, by using a shell, the user has to add the knowledge in the form of rules, and then provide relevant data to solve a problem.

1.3 Structure of Rule-Based Expert Systems

A rule-based expert system has five components:

1) The **knowledge base** that contains the domain knowledge represented as a set of IF-THEN production rules. The knowledge base is analogue to the long-term human memory. There is a total order among production rules. We can consider that each production rule has attached a priority that can be changed.

2) The **facts base** contains facts used to match against the IF part (i.e. the condition part) of rules stored in the knowledge base. This facts base is analogue to the short-term human memory (i.e. is very mobile).

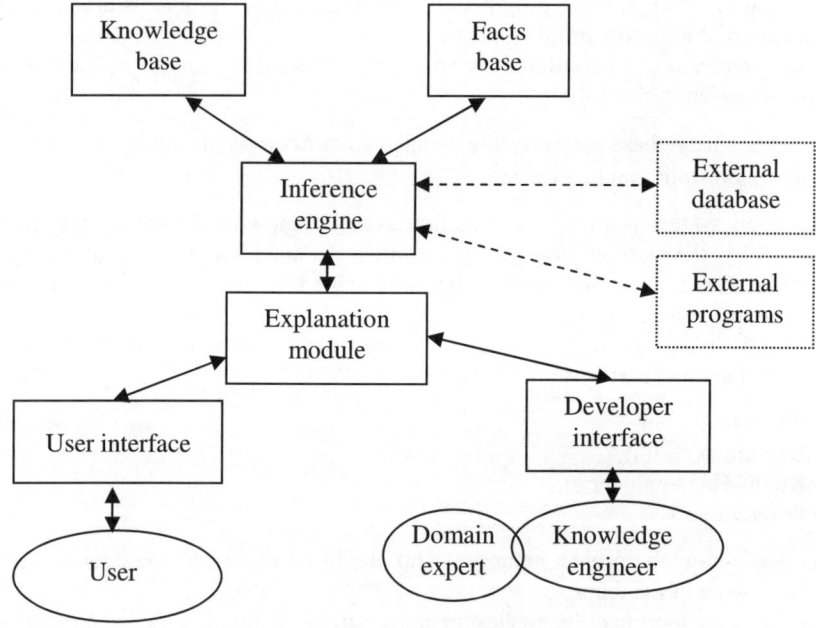

Fig. 1.1. Structure of an expert system

3) The **inference engine** carries out the reasoning by linking the rules to the facts, and by deducing new facts.

4) The **user interface** is the means of communication between users and the expert system.

5) The **explanation module** enables the user to ask the expert system how a particular conclusion is reached, and why a specific fact is needed. It is an essential component for developing activities.

4') The **developer interface** is needed to modify the knowledge base, and to store the knowledge in an external DBMS (database management system). Usually this interface includes knowledge base editor, debugging aids, and input/output facilities.

In the Figure 1.1 above a sketch of the medical MYCIN expert system is presented (as it was conceived by its developers).

1.4   Reasoning in an Expert System

From a logical point of view, the most important part of an expert system is the inference engine.

It works in a cyclic manner. At each stage, it takes a production rule – from the knowledge base – and compares it with facts stored in the facts base. This comparison occurs systematically either between the antecedent of the rule and the stored facts, or between the consequent and the stored facts. Therefore two different situations: **forward chaining** res. **backward chaining** are identified.

In Figure 1.2 a graphical representation of the two situations is presented.

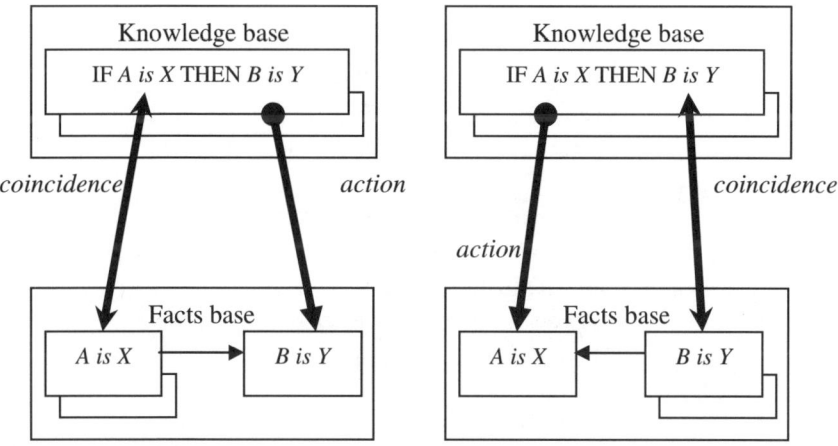

Fig. 1.2. Forward chaining vs. backward chaining

To simplify the presentation, let us suppose that all rules in the knowledge base are of single-fact consequent, i.e. they are **Horn**[1] **clauses**

IF $a_1 \wedge ... \wedge a_n$ THEN c

---

[1] Alfred Horn (1918-1988), American mathematician.

where a_1, ..., a_n and c are facts. The forward chaining reasoning is described as follows:

Step 0.   Start with a knowledge base K, a facts base F, and a fact f to be tested. Unmark all rules in K.

Step 1.   If all rules in K are marked, print "f cannot be inferred". Stop.

Step 2.   Select from K a rule ρ that is not marked. Compare to the facts in F all components of the antecedent of rule ρ.

> If all the components of the antecedent belong to the facts base F, then compare the tested fact f with the (unique) consequent of ρ.
>
>> If they coincide, then print "f is inferred". Stop.
>> If not, then
>>> Attach the consequent of ρ to the facts base.
>>>
>>> Mark the rule ρ as "fired".
>>>
>>> Unmark all rules previously marked as "visited".
>>> Continue with step 1.
>
> If not, then mark the rule ρ as "visited" and continue with step 1.

The backward chaining reasoning is described as follows:

Step 0.   Start with a knowledge base K, a facts base F, an empty stack S, and a fact f to be tested. Push f on stack S.

Step 1.   Check if the stack S is empty.
If yes, print "f cannot be inferred". Stop.
If not, extract (pop) t from stack S. Continue with step 2.

Step 2.   Check if there is an unmarked rule ρ in K whose consequent equals t.

> If yes, check whether all components of the antecedent of ρ are in F.
>
>> If yes, then
>>> Mark the rule ρ as "fired".
>>>
>>> Unmark all rules previously marked as "visited".
>>> Attach t to the facts base F.
>>> If t equals f, print "f is inferred by ρ". Stop.
>>>
>>> If not, continue with step 1.
>>
>> If not, then
>>> Push t on stack S.
>>> Push on S all components of the antecedent of ρ that are not in F.
>>> Mark rule ρ as "visited".
>>>
>>> Continue with step 2.
>
> If not, continue with step 1.

To explain in detail the functioning of the inference engine we will represent IF-THEN rules by the sign $\Rightarrow$, and the connector AND by the sign $\wedge$ as in classical propositional calculus. Facts will be represented by lowercase letters.

As an example, the knowledge base contains seven production rules:

rule 1:	$a \wedge r \wedge s \Rightarrow y$	rule 5:	$d \wedge e \Rightarrow s$
rule 2:	$b \Rightarrow p$	rule 6:	$p \Rightarrow x$
rule 3:	$c \wedge p \wedge q \Rightarrow r$	rule 7:	$s \Rightarrow z$
rule 4:	$d \Rightarrow q$		

Let us represent in Figure 1.3 the seven production rules above as the special arrows in a directed bicolor graph.

Let us point out that there is no variability in anyone of the seven production rules. Once a rule was used (i.e. "marked as fired") there is no need to fire it again!

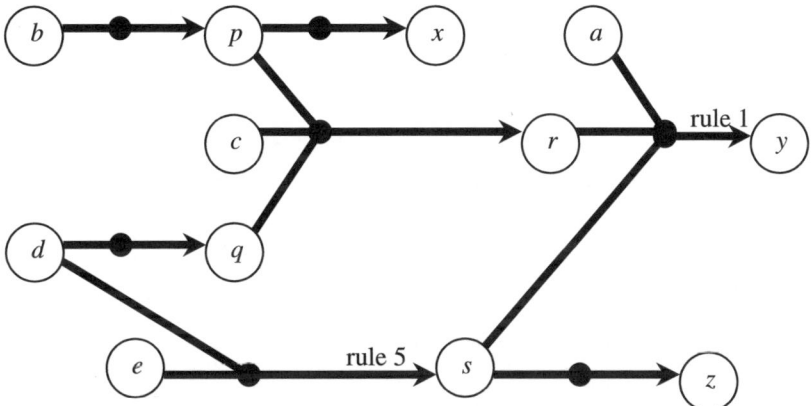

Fig. 1.3. Graph representation of the knowledge base

Suppose we "know" five facts a, b, c, d, e, and we want to establish, by forward chaining, whether z can be inferred or not from these facts.

Rule 1 cannot be fired at once, because we do not know yet the status of r, nor of s. The first rule to be fired is rule 2 (since its antecedent b is a fact), thus we attach p to the facts base.

Now two strategies are possible: either to continue from the fired rule downward, or to restart from the beginning. Both strategies give the same result: rule 4 should be fired next, and q should be attached to the facts base.

The two strategies above point now to different rules to be fired next: either rule 5, or rule 3. If we always continue from the fired rule downward (and we restart from the beginning if necessary) then z is obtained as a fact in five steps. However, by using the other strategy we need seven steps to obtain the result!

In Figure 1.4 below the first two steps are presented.

If we use backward chaining to check our "goal", i.e. to check whether z can be inferred or not, then a supplementary stack is needed. This stack will contain all "goals"; some of them are temporarily left aside because we reach for new "sub-goals" with a higher priority. The top of the stack will be compared to the THEN part of all rules. When it matches the THEN part of a rule, and all components of the corresponding antecedent (the IF part) are found in the facts base, then this top of the

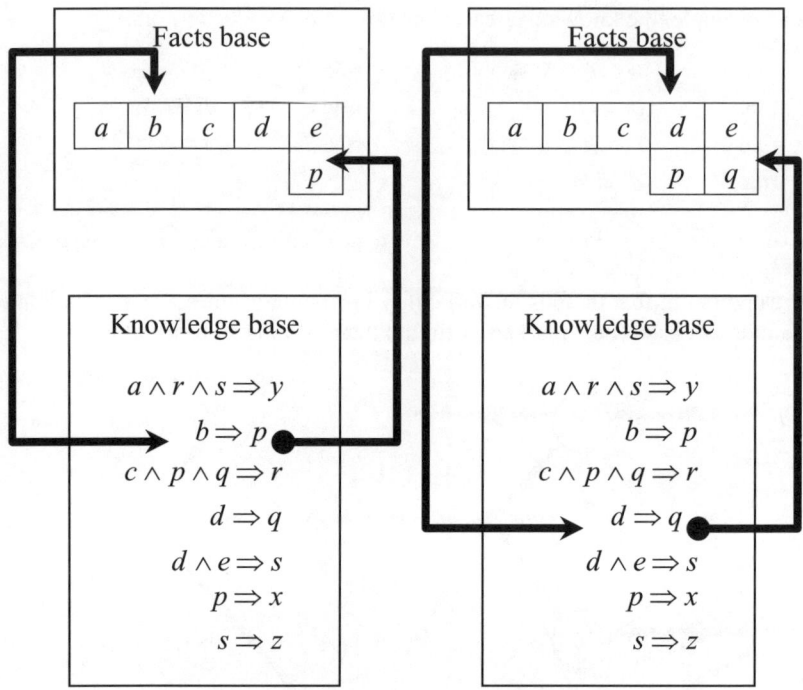

Fig. 1.4. First two steps in forward chaining example

stack will be removed and placed into the facts base. In all other situations, the components that are not found in the facts base will be pushed on the stack.

Our initial goal z will be the first top of the stack.

This time we start with "rule 7" because it is the first one to contain the fact z in the THEN part. In the IF part of this rule there is s, which is not – for the moment, in the facts base – and finding whether s can be inferred or not becomes the new sub-goal.

Due to rule 5, it is immediate that s is inferred; therefore it will be removed from the stack and placed into the facts base. Now by use of rule 7 z is inferred. These two steps are presented in Figure 1.5.

Most of medical expert systems (beginning with MYCIN, which was developed in 1975 to choose anti-bacterial therapy for patients suffering from a severe infection) are backward chaining driven.

Here is an example of production rules that are used in medical therapy:

rule 579:	IF	The infection that requires therapy is meningitis
	AND	The patient's chest X-ray is abnormal
	AND	Active-tb is one of the diseases that the patient's chest X-ray suggests
	THEN	there is strongly suggestive evidence that Mycobacterium-tb is one of the organisms (other than those seen on cultures or smears) that might be causing the infection

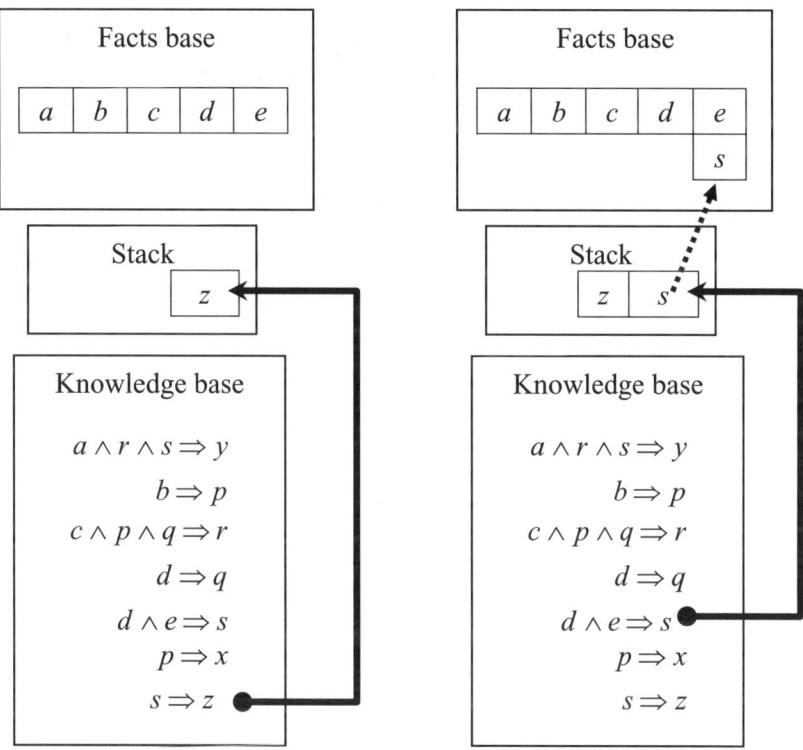

Fig. 1.5. First two steps in backward chaining example

and here is an example of how a production rule appears in MYCIN:

```
($AND (SAME CNTXT GRAM GRAMPOS)
      (SAME CNTXT MORPH COCCUS)
      (SAME CNTXT CONFORM CLUMPS))
(CONCLUDE CNTXT TALLY STAPHYLOCOCCUS TALLY 700)
```

1.5 Conflicts Resolution

Another task of the inference engine is to choose the production rule to be fired. Once fired, a rule may affect the firing of other rules. Let us consider a simple example:

rule 1:	IF	the traffic light is green
	THEN	go across the street
rule 2:	IF	the traffic light is red
	THEN	go across the street
rule 3:	IF	the traffic light is red
	THEN	stop and wait

It is clear that rule 2 and rule 3 are in conflict.

If the common IF part of rules 2 and 3 is true, then the inference engine has to choose to fire one of these rules. This is called **conflict resolution**.

It is obvious that the rules are ordered and the inference engine may choose to fire the first matching (rule 2 in our case). However, it is obvious also that is not always the best choice.

It is preferable to attach to each rule a priority. Usually the priorities are modified by the firing of a meta-rule.

If it has enough time, the inference engine can choose among all the rules and find the most specific one. This is called *the longest matching strategy*. Here it is an example:

rule 1:	IF	the season is autumn
	THEN	the advice is to take the umbrella
rule 2:	IF	the season is autumn
	AND	the sky is covered
	AND	the weather forecast is rain
	THEN	the advice is to stay at home

In this case the second rule will be chosen, its IF part being more specific.

Other conflict resolution method, called *data most recently entered,* is based on the time tags attached to each fact in the database.

Finally, we can have a *random choice*, based on a RANDOM function.

Thus, the methods of conflict resolution are:

– Natural order,
– The highest priority,
– The longest matching strategy,
– Data most recently entered,
– Random choice.

To improve the performance of an expert system we should supply the system with knowledge about knowledge (i.e. with meta-knowledge), which controls the production rules. Because we cannot distinguish formally between a rule and a meta-rule, the meta-rules will be placed on the top of the knowledge base.

In general, expert systems can deal with incomplete and uncertain data, and also permit inexact reasoning. However, the expert systems have three main shortcomings:

1) Opaque relations between rules,
2) Ineffective search strategy,
3) Inability to learn.

1.6 Solved Exercises

1)   The knowledge base is the following:

IF *A is true*, THEN *B is true*

IF *C is true*, THEN *D is true*
IF *E is true*, THEN *F is true*
IF *B is true* AND *D is true*, THEN *G is true*
IF *B is true* AND *F is true*, THEN *H is true*
IF *D is true* AND *F is true*, THEN *I is true*

and the facts base is the following:

A is true
E is true.

Using the methods of forward chaining and backward chaining, try to establish the goal *H is true*.

2) An expert system has the following knowledge base:

rule 1: $P \Rightarrow D$
rule 2: $P \Rightarrow E$
rule 3: $A \wedge B \Rightarrow D$
rule 4: $A \wedge B \Rightarrow G$
rule 5: $A \Rightarrow C$
rule 6: $B \Rightarrow P$
rule 7: $A \Rightarrow K$
rule 8: $D \wedge L \Rightarrow C$
rule 9: $D \wedge L \Rightarrow H$
rule 10: $X \Rightarrow A$
rule 11: $E \Rightarrow D$
rule 12: $A \Rightarrow L$
rule 13: $H \wedge W \Rightarrow Z$
rule 14: $P \wedge L \Rightarrow W$

(a) The facts base is the following:

B, C, X.

Show explicitly how our expert system will "prove" *H*, reasoning by backward chaining.

(b) If the expert system reasons by forward chaining and the facts base is the following

A, B,

which are the facts "proved" by the expert system?

3) The production rules are as follows

IF today_is_rain THEN tomorrow_is_rain
IF today_is_rain AND rainfall_is_low THEN tomorrow_is_dry
IF today_is_rain AND rainfall_is_low AND temperature_is_cold
        THEN tomorrow_is_dry
IF today_is_dry AND temperature_is_warm
        THEN tomorrow_is_dry
IF today_is_dry AND temperature_is_warm AND sky_is_overcast
        THEN tomorrow_is_rain

Given that each day can be only rainy or dry, forecast by backward chaining the weather tomorrow.

Using forward chaining, forecast the weather knowing that today_is_dry, temperature_is_warm, sky_is_overcast are facts.

4) The knowledge base is:

rule 1: $HA \Rightarrow MA$

rule 2: $GM \Rightarrow MA$

rule 3: $FE \Rightarrow BI$

rule 4: $(\neg CF) \wedge LE \Rightarrow BI$

rule 5: $EM \Rightarrow CA$

rule 6: $PT \wedge CL \wedge FO \Rightarrow CA$

rule 7: $MA \wedge HO \Rightarrow UN$

rule 8: $MA \wedge CC \Rightarrow UN$

rule 9: $MA \wedge CA \wedge TC \wedge DS \Rightarrow CH$

rule 10: $MA \wedge CA \wedge TC \wedge BS \Rightarrow TI$

rule 11: $DS \wedge LL \wedge LN \wedge UN \Rightarrow GI$

rule 12: $BS \wedge UN \Rightarrow ZE$

rule 13: $LN \wedge BI \wedge CF \wedge BW \Rightarrow OS$

rule 14: $CF \wedge BI \wedge BW \wedge SW \Rightarrow PI$

rule 15: $BI \wedge FW \Rightarrow AL$

and the known facts are: $\neg BS$, CF, DS, EM, $\neg FE$, HA, $\neg HO$, TC. (Of course, sign $\neg$ means "not".)

Try to "prove" CH, reasoning by backward chaining.

Solutions. 1) When using forward chaining, the fired rules are, in order, the 1st, the 3rd, and the 5th. The goal is obtained, when using backward chaining, by the same rules, fired in reverse order.

2) (a) H is "proved" after firing the rules 9, 12, 10, 1, and 6.

(b) All letters, except X, are "proved facts" by forward chaining, and all rules, except the 10th, are fired.

3) The result of backward chaining depends on known facts about weather today. Of course, using forward chaining, from the 4th and 5th rules we obtain as forecast tomorrow_is_dry and tomorrow_is_rain.

(The apparent contradiction of conclusions of the first two rules shows that the classical logic is not an adequate tool in expert systems. In Chapter 7 it will be shown how production rules are really used in classical expert systems.)

4) It is a useful example of "guessing" that cheetah (CH) is the animal, knowing that:

BS = has_black_stripes, CF = can't_fly, DS = has_dark_spots,
EM = eats_meat, FE = has_feathers, HA = has_hair, HO = has_hoofs,
and TC = has_tawny_color.

The other facts appearing in the rules may be interpreted as:

AL = is_an_albatros, BI = is_bird, BW = is_black_and_white,
CA = is_carnivore, CC = chews_cud, CL = has_claws,
DS = has_dark_spots, FO = has_forward_eyes, FW = flies_well,
GI = is_a_giraffe, GM = gives_milk, LE = lies_eggs,
LL = has_long_legs, LL = has_long_neck, MA = is_mammal,
OS = is_an_ostrich, PI = is_a_pinguin, PT = has_pointed_teeth,
SW = swims, TI = is_a_tiger, UN = is_ungulate, ZE = is_a_zebra.

2 Knowledge Representation

2.1 Data, Information and Knowledge

In Figure 2.1 the variations (in voltage) of the current at the extremity of a transmission wire is presented. A well-tuned device, which is able to eliminate noise, "captures" the current as the sequence of bits:

10101100101011111000010110001111001010

and prints it.

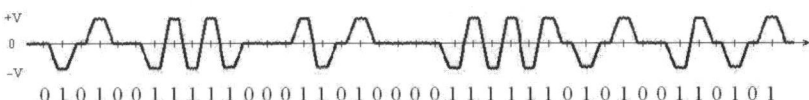

Fig. 2.1. Current in a point of a transmission wire

Suppose we possess the following knowledge: every bit should be complemented, then we have to group the bits in sequences of 8:

01010011 01010000 00111101 00111000 00110101

Finally, we have to interpret these sequences according to the ASCII code. The result is printed as
SP=85
Both the sequence of 40 bits and the sequence of 5 characters are considered as *data*.
Now, suppose Your brain is trained such as to interpret the sequence above as
"systolic pressure of the patient is 85"
which is an *information*.
All this helps us to understand the hierarchy of terms:

0. Variations of state of an object
1. Data
2. Information
3. Knowledge
4. Meta-knowledge.

E. Roventa and T. Spircu: Management of Knowledge Imperfection, STUDFUZZ 227, pp. 13–30.
springerlink.com   © Springer-Verlag Berlin Heidelberg 2009

We may possess knowledge that allows us to transform data. Certain knowledge allows us to extract information from data. Information means interpreted data. (How knowledge is acquired is not our concern.).

The meta-knowledge is knowledge about the knowledge that we have to use in a given situation.

In the previous chapter knowledge was represented in form of production rules. However, knowledge may be represented in several other ways. In the following sections we explore other possibilities: more complex logical systems, semantic nets, frames.

2.2 Logical Systems

Perhaps one of the oldest ways to represent knowledge is encountered in classical and predicate logic.

Today's logic is a vast area of human knowledge, dealing with all aspects of reasoning processes. Its history is ancient, more than 2000 years ago Aristotle (384-322 b.Ch.) described several simple inference rules. The last century witnessed, long before the appearance of computers, a tremendous expansion of different branches of logic.

Classical (i.e. Aristotelian) logic is a tool indispensable for all kind of human knowledge representation; moreover, it is the first auxiliary in formalizing arguments. We will present the elements of this logic and some of its extensions.

To construct a logical system we need:

1) A language,

2) A truth structure, which can be described either syntactically, or semantically, and

3) A list of inference rules.

The simplest example of a logical system is the propositional calculus. Its language is a part of the natural language (in our case, the English), more precisely that part formed by declarative sentences having a well defined truth value (either "true", or "false").

In inter-human communication processes information is transmitted, orders are given, questions are asked, and exclamations are made. The sentences having an imperative, interrogative or exclamatory component are left outside propositional calculus. Only pure declarative sentences belong to the propositional calculus. Examples:

"Where are you going?"   NO (it is interrogative)
"Give me the surgical knife!"   NO (it is imperative)
"Look how ill that person is!"   NO (it is exclamatory)
"Number five is odd"   YES (it is "true")
"A patient has fever"   NO (it is an imprecise sentence)
"John Johnson has fever"   YES (suppose we know it is "true")
"John Johnson is forty-five years old and has fever"
          YES (and it is a composed sentence)
"It rains now over Toronto"
          NO (because the truth value is not certain).

It is difficult enough to define precisely the term "proposition". For the sake of simplicity let us adopt the naïve approach: a proposition is simply a statement, expressed in a sequence of words, which has a precise meaning and is evaluated by Your brain either as "true" or as "false".

For example, we could accept that

"The result of sputum lab test is not good"

"Diabetes is a frequent chronic disease"

and

"The patient has tuberculosis"

are propositions. On the contrary,

"My patient has" and "The result of the lab test will be good"

are not propositions: the first has no meaning in itself (except perhaps in a specific context), the second cannot be evaluated now, it is "yet undecided".

Notice that

"The result of sputum lab test is <u>not</u> good"

has the same meaning as

"It is not true that the result of sputum lab test is good".

Also,

"Diabetes is a frequent chronic disease"

could be rewritten as

"Diabetes is a frequent disease <u>and</u> diabetes is a chronic disease".

It is possible to express any compound proposition in terms of simpler ones, using the so-called **logical connectors**.

There are four logical connectors used in inter-human communication:

1) "and", denoted by $\wedge$, used to form conjunction,

2) "or", denoted by $\vee$, used to form disjunction,

3) "if-then", denoted by $\Rightarrow$, used to form implication,

4) "if and only if", denoted by $\Leftrightarrow$, used to form equivalence.

A fifth sign $\neg$, read as "not", is used to form negations.

A proposition is called **atomic** if it cannot be expressed, using logical connectors $\wedge$, $\vee$, $\Rightarrow$ or $\Leftrightarrow$, in terms of simpler ones. For example, "The result of sputum lab test is not good" and "The patient has tuberculosis" are considered atomic propositions.

Atomic propositions that do not contain the sign $\neg$ are called simple.

However, there is a large gap between propositions above and formal propositions treated in propositional calculus. To be precise, let us specify that the language L of propositional calculus is based on the following alphabet

all lowercase Courier letters, _, $\neg$, $\wedge$, $\vee$, $\Rightarrow$, $\Leftrightarrow$, (,).

The underscore sign is used to replace the white spaces between words. Thus, propositions became words.

The "correct words" of the language L known as the **well-formed formulas**, are described by the following rules:

(R0) Any (finite) word composed by lowercase Courier letters and/or signs _ is a well-formed formula (w. f. f.)

(R1) $\neg\phi$ is a w. f. f. where ϕ represents a w. f. f.

(R2) $\phi \wedge \psi$, $\phi \vee \psi$, $\phi \Rightarrow \psi$, $\phi \Leftrightarrow \psi$ are w. f. f.'s where ϕ, ψ represent w. f. f.'s (not necessarily distinct)

(R3) (ϕ) is a w. f. f. where ϕ represents a w. f. f.

(R4) No other w. f. f. does exist.

It is customary to represent propositions (not necessary simple) by lowercase italic letters $p, q, r, \ldots$. For example,

`diabetes_is_a_frequent_chronic_disease`

is represented formally as

$p \wedge q$

where p represents `diabetes_is_a_frequent_disease` and q represents `diabetes_is_a_chronic_disease`".

In general, if p and q represent propositions, then:

$\neg p$ represents the proposition interpreted as "It is not true that p ",

$p \wedge q$ represents the proposition interpreted as " p and q ",

$p \vee q$ represents the proposition interpreted as " p or q ",

$p \Rightarrow q$ represents the proposition interpreted as "if p , then q ",

$p \Leftrightarrow q$ represents the proposition interpreted as " p if and only if q ".

The main objective of propositional calculus is to establish the truth-value of compound propositions given the truth-values of simple components. The main rules used are summarized in the following two tables:

p	$\neg p$
true	false
false	true

p	q	$p \wedge q$	$p \vee q$	$p \Rightarrow q$	$p \Leftrightarrow q$
true	true	true	true	true	true
true	false	false	true	false	false
false	true	false	true	true	false
false	false	false	false	true	true

Notice that conjunction $p \wedge q$ is "true" only if both components p and q are "true".

Notice also the situation of the implication: if the "premise" p is "false", then $p \Rightarrow q$ is "true" regardless of the truth-value of the "conclusion" q .

As for the equivalence $p \Leftrightarrow q$, notice that it is true whenever p and q have the same truth-value.

Sometimes we call **fact** a "true" atomic proposition, and **knowledge** a "true" implication (i.e. a proposition of the form $p \Rightarrow q$ that is "true").

Several syntactic rules ("laws") of propositional calculus exist.

(I) Law of double negation

$\neg(\neg p)$ has the same truth-value as p

Hence, if it is "false" that "The result of sputum lab test is not good", then it is "true" that "The result of sputum lab test is good".

(II) Law of contraction

$p \wedge p$ has the same truth-value as p

$p \vee p$ has the same truth-value as p .

(III) Law of commutation (not valid for the implication!)

$p \wedge q$ has the same truth-value as $q \wedge p$ (whatever the truth-values of p
    and of q)

$p \vee q$ has the same truth-value as $q \vee p$

$p \Leftrightarrow q$ has the same truth-value as $q \Leftrightarrow p$.

(IV) Law of association

$(p \wedge q) \wedge r$ has the same truth-value as $p \wedge (q \wedge r)$

$(p \vee q) \vee r$ has the same truth-value as $p \vee (q \vee r)$.

(V) Laws of distribution (of conjunction with respect to disjunction, and of disjunction w. r. t. conjunction)

$p \wedge (q \vee r)$ has the same truth-value as $(p \wedge q) \vee (p \wedge r)$

$p \vee (q \wedge r)$ has the same truth-value as $(p \vee q) \wedge (p \vee r)$.

(VI) Laws of De Morgan (duality of conjunction and disjunction)

$\neg(p \wedge q)$ has the same truth-value as $(\neg p) \vee (\neg q)$

$\neg(p \vee q)$ has the same truth-value as $(\neg p) \wedge (\neg q)$.

Hence according to the first law of De Morgan, the negation of "Diabetes is a frequent chronic disease" could be expressed as "Diabetes is not a frequent disease or diabetes is not a chronic disease".

(VII) Law of the implication

$p \Rightarrow q$ has the same truth-value as $(\neg p) \vee q$.

According to this last law, the proposition "If the result of sputum lab test is not good, then the patient has tuberculosis" has the same truth-value as the proposition "The result of sputum lab test is good, or the patient has tuberculosis".

(VIII) Law of the equivalence

$p \Leftrightarrow q$ has the same truth-value as $(p \Rightarrow q) \wedge (q \Rightarrow p)$, or as

$$(p \wedge q) \vee ((\neg p) \wedge (\neg q)).$$

Although humans use in communication all five logical signs, it is clear from the above laws that some of these are superfluous. Clearly $\Rightarrow$ and $\Leftrightarrow$ are expressed in terms of $\neg$, $\wedge$ and $\vee$, as the laws (VII) and (VIII.2) above show.

A proposition is said **to be in normal form** if it is expressed as a disjunction

$$p_1 \vee p_2 \vee ... \vee p_n$$

where each component p_i is expressed as a conjunction

$$q_{i1} \wedge q_{i2} \wedge ...$$

where each component q_{ij} is an atomic proposition.

The results above show that every compound proposition is equivalent to a normal form. This allows a certain "automatic" treatment of the propositions in propositional calculus.

Let us remind the **Horn clauses**, which are compound propositions of the form

$$(p_1 \wedge p_2 \wedge ... \wedge p_n) \Rightarrow q$$

i.e. implications having as "premise" a conjunction of several propositions. Interpreted in the following way: "q is true if we are able to establish that $p_1, p_2, ..., p_n$ are all true", the Horn clauses constitute the basis of a logic-dedicated programming language, whose name is Prolog. This language is well adapted to the so-called Artificial Intelligence.

Combining the law of double negation (I) with the law of De Morgan (formula VI.1 above), we obtain

$p \wedge q$ has the same truth-value as $\neg((\neg p) \vee (\neg q))$

showing that $\wedge$ is expressed in terms of $\neg$ and $\vee$. Hence we could reduce in propositional calculus, without loosing any truth-value, the set of logical signs to $\neg$ and only.

We could reduce the set of logical signs in another way. Namely, reversing the law of the implication (VII) and taking into account the law of double negation (I), we obtain the formula

$p \vee q$ has the same truth-value as $(\neg p) \Rightarrow q$

showing that $\vee$ is expressed in terms of $\neg$ and $\Rightarrow$. It is easy to establish that

$p \wedge q$ has the same truth-value as $\neg(p \Rightarrow (\neg q))$

and the law of the equivalence (VIII.1) shows that all five logical signs are expressed in terms of $\neg$ and $\Rightarrow$.

An obvious question appears. Could the set of logical signs be reduced to only one?

The answer is negative if we restrain to these five signs above. However, the answer is positive if we consider other logical signs!

Indeed, there exist other logical connectors. In some circumstances humans use the "either-or" connector, denoted by XOR , described by the following truth table:

p	q	$pXORq$
true	true	false
true	false	true
false	true	true
false	false	false

It is obvious that $pXORq$ has the same truth-value as $\neg(p \Leftrightarrow q)$. The normal form equivalent to $pXORq$ is $(p \wedge (\neg q)) \vee ((\neg p) \wedge q)$ and the most important property of this connector is the following

$(pXORq)XORq$ has the same truth-value as p .

Perhaps the most useful "non-human" logical connector is the so-called $NAND$, whose truth table is the following

p	q	$pNANDq$
true	true	false
true	false	true
false	true	true
false	false	true

For this connector it is easy to establish that

$\neg p$ has the same truth-value as $pNANDp$,

$p \Rightarrow q$ has the same truth-value as $pNAND(pNANDq)$

therefore $NAND$ alone is self-sufficient to express all human logical signs! Of crucial importance is the fact that, by using p-n-p transistors, simple electronic circuits simulating $NAND$ operation can be assembled. These serve as "bricks" for building the arithmetic-logic unit of every contemporary computer.

2.3 Predicate Calculus

In propositional calculus the sentences admit decompositions in "simple" sentences (that do not contain logical signs $\neg$, $\wedge$, $\vee$, $\Rightarrow$, $\Leftrightarrow$). Parentheses are also used, though they are not necessary. For example, to avoid all ambiguity in the expression $\neg p \vee q$, we could insert parentheses specifying either as $(\neg p) \vee q$, or as $\neg(p \vee q)$.

The language of propositional calculus is very much simplified in contrast to the natural language (English). For example, in English $\wedge$ means not only "and", it means also "but" or "and then". Temporal aspects are not taken into account. However, the main deficiency of the propositional calculus is the impossibility to treat general sentences.

A first extension, much more comprehensive, is the predicate calculus. In the frame of predicate calculus, the sentences admit a decomposing into:

– Consants,
– Variables,
– Functions,
– Predicates,
– Logical signs (the same as in the propositional calculus), and
– Quantifiers.

Constants name objects. Variables denote generically objects from a domain. Functions may have arbitrary values and one or several arguments, which are variables. Predicates are special functions, their specificity consisting of the values they may take, namely only T ("true") or F ("false"). The arguments of a predicate are either constants, or variables. The universal quantifier is denoted $\forall$ ("any"), and the existential quantifier is denoted $\exists$ ("exists"). The quantifiers do not appear alone; any quantifier always precedes a variable. (Notice that in predicate calculus quantifying predicates is not allowed!)

Let us consider the predicate (of a single variable X, which denotes generically persons)

`is_a_physician(X)`

Starting from this predicate, the following simple sentence is built

`is_a_physician(john)`

where john is a constant. Moreover, the following two sentences, more complex, are built:

$\forall$X(`is_a_physician(X)`) meaning "any person is a physician",

$\exists$X(`is_a_physician(X)`) meaning "there exists physician(s)".

If the domain of variable X is the persons living in a big city, the first sentence is false and the second is true.

Let us consider the example of a predicate π of two variables X, Y, both variables denoting generically persons from a domain. The expression:

$\pi(X,Y)$

may be interpreted as "X is the son of Y" or, what is the same, "Y is the father of X".

Starting from this predicate several particular sentences are built, such as:

π(john,peter) i.e. "`john_is_the_son_of_peter`".

Moreover, four sentences involving one quantifier are assembled:

$\forall$X($\pi(X,Y)$)   "is a universal father"
$\forall$Y($\pi(X,Y)$)   "is a universal son"
$\exists$X($\pi(X,Y)$)   "is a father"
$\exists$Y($\pi(X,Y)$)   "is a son".

In these four examples one variable is "bound", the other is "free"; this means our four sentences are in fact predicates of a single variable! Further examples (in which two quantifiers appear):

$\forall X (\forall Y (\pi(X,Y)))$ "any son has as father any person",

$\forall X (\exists Y (\pi(X,Y)))$ "any son has at least one father",

$\exists X (\forall Y (\pi(X,Y)))$ "there exists at least one son, such that any person is his father",

$\exists X (\exists Y (\pi(X,Y)))$ "there exists at least a son that has at least one father",

$\forall Y (\forall X (\pi(X,Y)))$ "any father has any person as son",

$\forall Y (\exists X (\pi(X,Y)))$ "any father has at least one son",

$\exists Y (\forall X (\pi(X,Y)))$ "there exists at least one father having any person as a son",

$\exists Y (\exists X (\pi(X,Y)))$ "there exists at least a father that has at least one son".

It is easy to find out the truth-values of any sentence above. Moreover, it is easy to detect the logical equivalence between $\forall X \forall Y$ and $\forall Y \forall X$, res. between $\exists X \exists Y$ and $\exists Y \exists X$.

In predicate calculus the (English) sentences "John is a physician" and "any patient is treated by a physician" are expressed, by the intermediate of the predicates is_a_physician(), is_a_patient() and treats(), as follows:

is_a_physician(john)

$\forall X (is\_a\_patient(X) \wedge \exists Y (is\_a\_physician(Y) \wedge treats(Y,X)))$

(However, there are differences between meanings in the natural language and in predicate calculus. In natural language "any patient is treated by a physician" normally means "any particular patient is treated by a single physician". The formal expression of the latter meaning is more complicated.)

Of course, any sentence from the predicate calculus has a well-defined truth-value, which is either T ("true"), or F ("false"). All the "laws" of propositional calculus are still valid in the predicate calculus. Several new "laws" are valid here, such as:

$\neg \exists X (\phi(X))$ has the same truth-value as $\forall X (\neg(\phi(X)))$

which expresses the logical equivalence between sentences "there is no ..." and sentences in which the predicate is negated ("for any ..., not ...").

However, in general the logical truth-value of a "correct formula" – obtained by quantifying the variables that appear in a predicate – depends essentially on the interpretation of the predicate, in fact on the interpretation of all components of the "formula".

If we want to formalize the predicate calculus, so as to avoid any possible paradox, a good start is a clear definition of an "atomic formula", followed by a good recursive definition, based on clear rules, of what a "correct formula" should be. This is not the intent of this book.

Very important are the formulas of type "implication", i.e. of the form $h \Rightarrow c$ ("if hypothesis h, then conclusion c") which extend the production rules. An example, in which only one predicate π of two variables is involved, is the following formula:

$\pi(X,Y) \Rightarrow \exists Z (\pi(X,Z) \wedge \pi(Z,Y)).$

Let us present two different interpretations. First, the domain is formed by persons, and $\pi(X,Y)$ is interpreted as "X is the son of Y" (as above). The formula is "true" when the "left hand" $\pi(X,Y)$ is "false". However, it is immediate that in this interpretation the "right hand" formula $\exists Z(\pi(X,Z) \wedge \pi(Z,Y))$ is "false" when $\pi(X,Y)$ is "true". Because the implication "true"$\Rightarrow$"false" has a "false" truth-value, it is clear that the sentence above may be "false".

Now let the domain be the set of real numbers (i.e. the variables X, Y take real values), and $\pi(X,Y)$ is interpreted as "$X < Y$". This time the "right hand" formula $\exists Z(X<Z \wedge Z<Y)$ is "true" when $X < Y$ is "true". Because "true"$\Rightarrow$"true" is "true", our sentence above is always "true".

2.4 Inference Rules in Classical Logic

In any logic, the inference of truth follows some rules. Three types of inference are known:

- Deductive,
- Inductive,
- Abductive.

Deductive inference rules are known since Aristotle. Perhaps the best known is *modus ponens*, which in propositional calculus is formally expressed as:

$$\frac{p}{p \Rightarrow q} \mapsto q \text{ or as } \frac{p, p \Rightarrow q}{q}$$

and which is interpreted in natural language as: "knowing that a proposition p is true and that the implication $p \Rightarrow q$ is true, the truth of the proposition q is inferred".

This rule is "certain" and is easily extended to predicate calculus, as follows:

$$\frac{\phi(a)}{\forall X(\phi(X) \Rightarrow \psi(X))} \mapsto \psi(a)$$

and allows the deduction of some "true" conclusions such as, for example, "Socrates is mortal" from "Socrates is human" and "all humans are mortal".

Another very well known "certain" rule is the so-called *syllogismus*; here only implications appear:

$$\frac{p \Rightarrow q}{q \Rightarrow r} \mapsto p \Rightarrow r$$

This rule is interpreted as follows: knowing that implications $p \Rightarrow q$ and $q \Rightarrow r$ are true, the truth of the implication $p \Rightarrow r$ is inferred.

In logical software another "certain" rule is implemented. This is the so-called **resolution**, which is presented as follows:

$$\frac{p \vee q}{(\neg q) \vee r} \mapsto p \vee r$$

and allows an automatically treatment of the list containing all known true facts.

Inductive inference rules allow generalizing particular facts. For example, from true facts "John is ill", "Peter is ill", "George is ill" we infer by induction that "everybody is ill". We feel that inductive inference rules are not certain. Of course, there is an exception: the mathematical induction.

The **abductive** inference rule:

$$\frac{p \Rightarrow q}{q} \mapsto p$$

is, of course, "uncertain". However, it is largely used in scientific research to identify the "reason" of a phenomenon when its effects are known.

Let us summarize. A logical system consists of the following:

1) A formal system, consisting of:
   - the syntax of the logical language (how to construct well-formed formulas)
   - the semantics of the logical language (how the formulas are in concordance with our manner of thinking about reality), and

2) A set of inference rules for deducing other formulas from the axioms.

There are an infinite number of possible logical systems. Two logical systems can be different because they use:

a) Different symbols in the alphabet,
b) Different rules for obtaining the well-formed formulas,
c) Different number of truth values,
d) Different manners for extending the truth values from elementary sentences to compound sentences;
e) Different sets of inference rules, or
f) Different sets of axioms.

A particular logical system could be used to represent knowledge.

2.5 Semantic Nets

Semantic nets are another type of classical techniques for representing information and knowledge. In fact, from the mathematical point of view, semantic nets are directed (labeled) graphs. The vertices (nodes) of such a graph represent objects, real or abstract. The edges (arrows, links) are used to express relationship between objects.

The links provide the basic structure for organizing knowledge.

Look at the example in Figure 2.2 below. It is easy to infer that Ann and Bill are the grandparents of John (even if there is no explicit links "grandfather_of", "grandmother_of" and not all the knowledge is represented by links, see for example the case of "sister_of"). This net may cover also the explanation for John's disability!

As another example, "virus", "sickness" and "syndrome" are vertices, and "may_cause" is an arrow from "virus" to "sickness" (or to "syndrome"). The usual interpretations of arrows in medicine are the following: "is_a_part_of", "is_a_propriety_of", "treat", "determine" etc. Usually arcs appear in pairs, for example "treat" and "is_treated_by", "determine" and "is_determined_by", "substance_measured" and "measured_by".

However, the use in medicine of semantic network is limited by the complexity of graphical representation of real medical problems.

The most used links in general semantic nets are expressed by:

- is_a, which means "is an instance of", i.e. "belongs to a class", and
- a_kind_of, which means subordination between classes. (In fact, between a class and a super-class).

Of course, the objects in a class have attributes in common: each attribute has a value. The combination attribute-value is called **property**. Thus

property(Attribute,Value)

is a special predicate. Triples

object-attribute-value

are used to characterize all the knowledge in a semantic net.

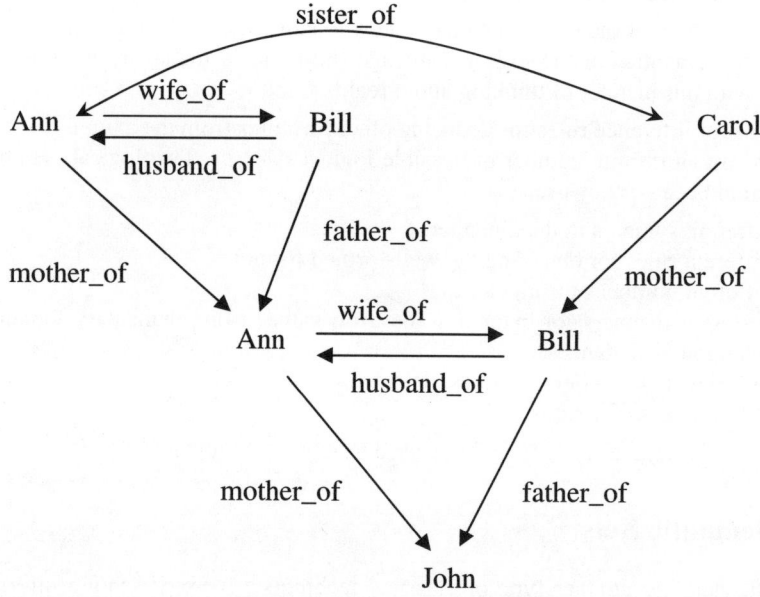

Fig. 2.2. Example of a semantic net

2.6 Frames

When humans face a new situation, they adapt the closest "frame" which they found in their memory to represent that situation. It is by comparison with the prototypes that adaptation takes place. The description of a universe is achieved gradually by increasing the experience regarding that universe.

To represent knowledge, in [Minsky 1975] it was proposed to adopt frames, i.e. abstractions in which objects are classified taking into account their general properties and first the most important ones. The frames constitute a manner of implementing proto-types. By using frames, we can reason without paying attention to the irrelevant details.

If the semantic networks are represented by bi-dimensional graphs, then in the representation of frames we add a third dimension. More precisely, each vertex (node) can have a more or less complex structure.

The frames are structures that present limited aspects of knowledge. To give a general definition, we consider that a **frame** is a name that names a group of **slots** and the values that fulfill the attached **fillers**. (Of course, the frame name could be considered as the value of filler attached to a particular slot NAME.)

Every filler has its own type; the types are extremely diversified, from the Boolean type to the type of the frame itself.

We have to stress that each filler is fulfilled, from the frame creation, which an implicit value. We can fulfill the fillers, following their importance, by replacing the implicit value by a significant value

Consider, as an example, the frame that represents the notion of car.

(NAME:) *Car*
make:
model:
year: 2006
speed: 5
engine:
tires: **4**
color:
...

Some fillers are empty; others contain implicit (by default) values such as:

tires: 4

The utility of frames lies in hierarchical frame systems and inheritance of properties; a frame can be the value for a filler of another frame, thus the inheritance.

Here is a simple example of inheritance:

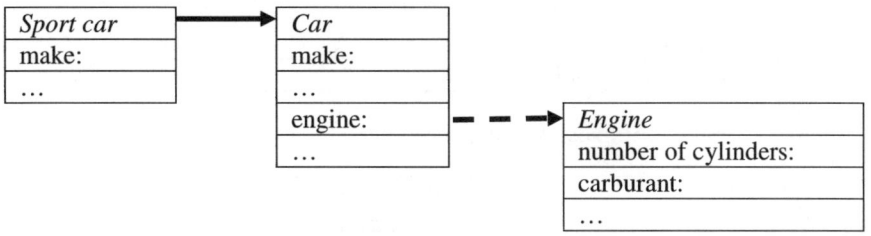

The solid arrows represent relations between a frame and a super-frame. The interrupted arrows represent relations between a filler and a value that is a frame.

It has to be stressed that values of fillers could be also logical procedures or even computer programs.

A frame description can be expressed in terms of a rule system.

However, it is not a good idea to introduce predicates for each possible filler (frame); it is better to introduce predicates that express relationships in a frame system.

predicate:	means:
frame(X)	X is a frame
is_a(X,Y)	The frame X is a sub-frame of Y
slot(S,X)	S is a filler of frame X
default(V,S,X)	V is the implicit value of filler S from frame X

The following rule

$$is\_a(X,Y) \wedge is\_a(Y,Z) \Rightarrow is\_a(X,Z).$$

expresses the hierarchical subordination between the given frames.

High quality medical vocabularies, as for example *Medical Entities Dictionary*, use frames for knowledge representation. Here is an (incomplete!) example from MED:

Slots for frame *14 Anatomical Structure*

	Slot Name	Slot Value(s)
0	MED-CODE	14
1	UMLS-CODE	T017
2	NAME	ANATOMICAL STRUCTURE
3	DESCENDANT-OF	1 Medical Entity
4	SUBCLASS-OF	1 Medical Entity
5	SYNONYMS	ANATOMIC ENTITY
6	PRINT-NAME	Anatomical Structure
7	HAS-PARTS	
8	PART-OF	
11	DEFINITION	
37	SITE-OF-PROBLEM	
50	MAIN-MESH	
51	SUPPLEMENTARY-MESH	
175	OBSERVATION-SITE-OF	
237	DEFAULT-SHORT-DISPLAY-NAME	
238	DEFAULT-DISPLAY-NAME	

Another example of frame, taken from medicine:

Slot Name	Slot Value(s)
NAME:	Acute glomerulonephritis
Triggered_by:	Facial edema, …
Confirmed_by:	Asthenia, anorexia, …
Caused_by:	Recent streptococci infection
Causes:	…
Complications:	…
Differential diagnostic:	…

We notice that the fillers are used to specify values but also to specify the connection with others concepts (confirmed_by).

May be the most important structure of frame type is the one standardized in the **Arden syntax**. Here the fillers are grouped in three categories: a) maintenance, b) library, and c) knowledge. In the inside of each filler a different formalism can be used: production rules, detailed descriptions, program codes, etc. Because all knowledge has to be implemented on the computer, data concerning the version, the author, and the creation date is inserted in the maintenance fillers. In the library fillers the goal, the key words for index creation, the details explanation and the possible connections with other information are inserted. Finally, in the knowledge filler the data type, the logic and the action (eventually as computer programs in pseudo-code) are inserted.

Here is the structure of an Arden frame (the optional fillers are not underlined)

Maintenance slots:	Library slots:	Knowledge slots:
Title:	Purpose:	Type:
Filename:	Explanation:	Data:
Version:	Keywords:	Priority:
Institution:	Citations:	Evoke:
Author:	Links:	Logic:
Specialist:		Action:
Date:		Urgency:
Validation:		

Of course, the knowledge slots are the most important. In the filler attached to Data the data are placed. In the filler attached to Logic

- data from filler Data are used,
- tests are performed, and
- decision whether the filler Action should be used or not is taken.

We can place here (in filler Action) all kind of programs, starting with elementary computations until complex rules needed to classify an object.

2.7 Solved Exercises

1) Translate the following sentences into predicate language:

Basketball players are tall.
John sent the email to all his fiends.
John sent the email only to his friends.

2) Express in predicate calculus the following set of sentences:

John likes fruits.
Bananas are fruits.
People eat what they like.
Does John eat bananas?

3) (From [Luger 2002]) The following conceptual graph (which is not a semantic net!) represents the sentence

John cut down the tree using an axe.

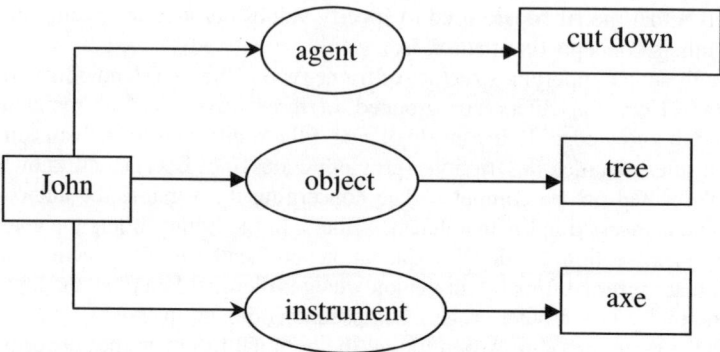

Express in natural language the knowledge represented in the graph:

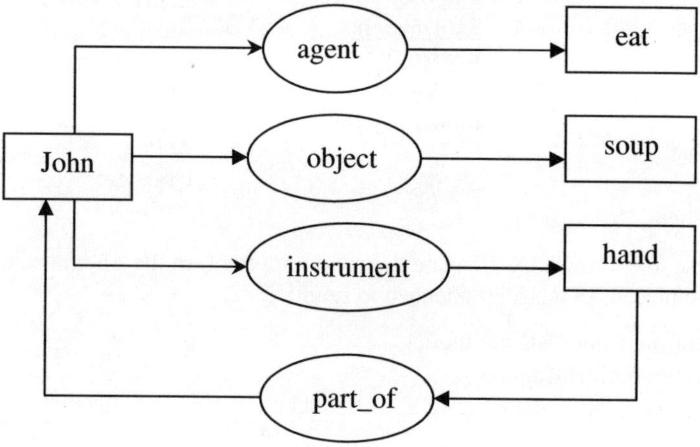

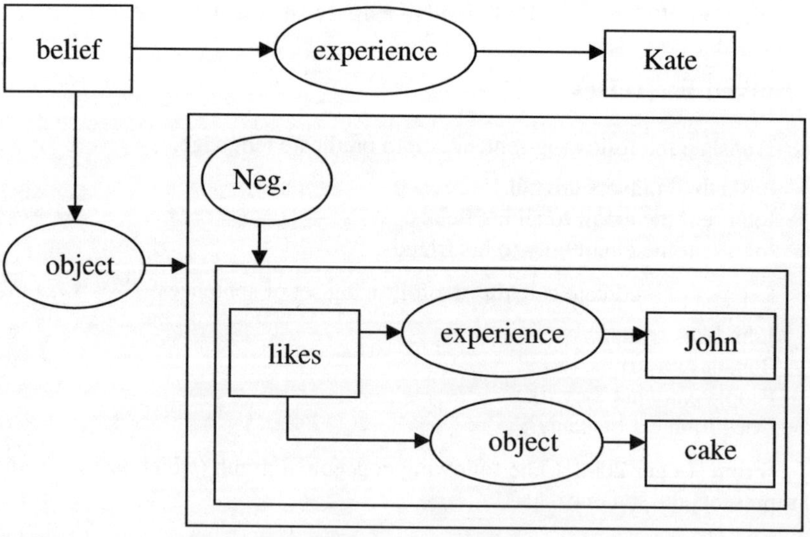

4)   (From [Rich and Knight 1991]) Enclose the following set of sentences into a formalism that facilitates the representation of knowledge they contain:

> When you go to a soccer game, you usually buy a ticket, hand the ticket to the ticket taker, and then go to find a seat.
> Sometimes you buy a hotdog before going to your seat.
> When the game is over you leave the arena.
> Alain went to the soccer game.
> Did Alain buy a ticket?

5)   Imagine a frame for the books of your library:

Solutions. 1) Obvious translations are:

$\forall X (\texttt{is\_player(X,basketball)} \Rightarrow \texttt{is\_tall(X))}$, or
$\forall X (\texttt{is\_basketball\_player(X)} \Rightarrow \texttt{is\_tall(X))}$,

$\forall Y (\texttt{friend(Y,john)} \Rightarrow \texttt{sent(john,Y,email))}$,

$\forall Y (\texttt{sent(john,Y,email)} \Leftrightarrow \texttt{friend(Y,john))}$.

2) The sentences are formalized in predicate calculus by using a constant $\texttt{john}$, four variables

$\texttt{F}$ denoting a generic fruit,

$\texttt{B}$ denoting a generic banana,

$\texttt{P}$ denoting a generic person (people),

$\texttt{O}$ denoting a generic object

and four predicates $\texttt{is\_fruit()}$, $\texttt{is\_banana()}$, $\texttt{likes()}$, $\texttt{eats()}$, as follows:

rule 1: $\forall F (\texttt{is\_fruit(F)} \Rightarrow \texttt{likes(john,F))}$

rule 2: $\forall B (\texttt{is\_banana(B)} \Rightarrow \texttt{is\_fruit(B))}$

rule 3: $\forall P \forall O (\texttt{likes(P,O)} \Rightarrow \texttt{eats(P,O))}$

If we express the question as follows
$\forall B (\texttt{is\_banana(B)} \wedge \texttt{eats(john,B))}$

its truth is easily obtained. Starting with a constant b such that
$\texttt{is\_banana(b)}$

is true, we obtain:

from rule 2: $\texttt{is\_fruit(b)}$ is true,

from rule 1: $\texttt{likes(john,b)}$ is true,

from rule 3: $\texttt{eats(john,b)}$ is true. Thus the answer to the question is "yes".

However, two problems appear:

a) It is supposed "john" is "people". Rule 3 should be expressed, more exactly, as:
rule 3': $\forall P \forall O (\texttt{is\_people(P)} \wedge \texttt{likes(P,O)} \Rightarrow \texttt{eats(P,O))}$
b) Our question was roughly expressed as
$\forall B (\texttt{is\_banana(B)} \wedge \texttt{eats(john,B))}$

In fact, this should be translated in natural language as "John eats all bananas (in the world)" or "John eats any banana (he grasps)". The use of two quantifiers ∀, ∃is an oversimplification!

Notice that in Prolog language all the above is expressed more clearly in four clauses and two facts:

```
likes(john,F) :- fruit().
fruit(F) :- banana(F).
eats(P,O) :- person(P),object(O),likes(P,O).
object(O) :- fruit(O).
person(john).
banana(b).
```

the goal being for example:

```
eats(john,b)?
```

3) "John eats soup using his hands", res. "Kate's belief is that John doesn't like cakes". These interpretations are obvious. Note the waste of space when using conceptual graph representations instead of sequences of characters.

4) At least two more quantifiers, namely "usually" and "sometimes", are necessary. There is not enough knowledge about Alain to precisely answer the question. An adequate answer could be "we are not sure".

People are "sure" that Alain left the arena. However, classical logic does not allow us to derive, directly or indirectly, the sentence "Alain left the arena" from the sentences we "know". (Of course, we do not know whether the game is over or not. It seems a lot of knowledge is "supposed".)

5) In recent books, the back of the title page contains several descriptors:

ISBN (10 and/or 13)
Edition number
Printing number

a.s.o. All these should be considered as slots. However, it is a good idea to respect the Arden syntax and to insert in our frame structure also maintenance and/or library slots. Especially very comprehensive <u>Keywords</u> slot would be extremely useful.

3 Uncertainty and Classical Theory of Probability

3.1 Taxonomy of Imperfection

Generally, a piece of evidence e is obtained after intecrpretation (in the "brain" of the observer) of a message received from one or several "sensors".

The piece of evidence e may come from different worlds $\omega, \omega_1, \omega_2, \ldots$ (in which e is "true") and the "brain" may have an opinion about who the actual world is.

We will follow here the approach of Philippe Smets [Smets 2000]. A piece of evidence e is called **perfect** if it is

1) Consistent (contradiction-free),
2) Precise (imprecision-free), and
3) Certain.

Unfortunately, the observer (or even the sensor) has to deal with all kind of imperfect data. Therefore, an imperfection label could be attached to the piece of evidence.

Many types of "error" could contribute to uncertainty. Let us present the most common, considering statements (propositions) that could be made about a valve intended to open/close a pipe, res. about a device that should measure the cardiac frequency (statements sent to physician *You* by a nurse):

Statement	Error	Reason
Turn the valve off	Ambiguity	What valve?
I measured the cardiac frequency		Which patient?
Turn valve#1	Incompleteness	Which way, on or off?
I measured the cardiac frequency of patient#1		It was good, or bad?
Turn valve#1 off	Incorrect	Correct is "on"
The cardiac frequency of patient#1 is good		In fact it is not good
Turn valve#1 to 5	Imprecise	Correct is 5.4
The cardiac frequency of patient#1 is 90		In fact is 92

E. Roventa and T. Spircu: Management of Knowledge Imperfection, STUDFUZZ 227, pp. 31–87.
springerlink.com   © Springer-Verlag Berlin Heidelberg 2009

Turn valve#1 to 5.4 The cardiac frequency of patient#1 is 92	Inaccurate	Correct is 9.2 In fact is 68
Turn valve#1 to 5.4 or 8.2 The cardiac frequency of patient#1 is 92 or 68	Unreliable	Impossible several simultaneous values
Valve#1 setting is at 7.3 The device shows the cardiac frequency of patient#1 is 73	Bias	Should be 7.5 It should be 75; the device is deteriorated
Valve#1 setting is at 5.4 or 5.5 or 5.3 The device shows the cardiac frequency of patient#1 fluctuant around 73	Random	Statistical fluctuation The fluctuation is random?
Valve#1 is not stuck The cardiac frequency of patient#1 is not constant	False negative	Valve is stuck In fact it is constant
Valve#1 is not stuck because it's never been stuck before The cardiac frequency of patient#1 is not constant because it was never constant before	Invalid induction	Valve is stuck In fact is constant now
Output is normal and so valve#1 is in good condition The device shows normal hence the cardiac frequency of patient#1 is good	Invalid deduction	Valve is stuck in "open" position The device is defective

Considering that our piece of evidence is a statement, the following taxonomy (classification) of imperfection is identified by Smets.

A) Inconsistency, when no world satisfies the statement (in the opinion of the observer). This can be:

A1) Conflicting (disagreement among the data),

A2) Incoherent (conclusions drawn from data, which does not make sense),

A3) Confused (a milder form of incoherence).

B) Imprecision, when several worlds satisfy the statement. Two major ramifications:

B1) Data without error

B1a) Missing (incomplete, deficient),

B1b) Vague (approximate, ambiguous, fuzzy);

B2) Data with error(s)

B2a) Erroneous (just wrong),

B2b) Inaccurate (essentially imprecise; however, not completely erroneous),

B2c) Invalid (would lead to unacceptable conclusions),

B2d) Distorted (wrong, but not far from correct),

B2e) Biased (systematic error),

B2f) Meaningless (cannot be fitted to reality).

C) Uncertainty, which depends on the quality of the evidence and/or the quality of the "brain" of the observer. Two major ramifications:

C1) Objective (related to the evidence)

C1a) Propensity: Random (subject to change), Likely (will probable occur),

C1b) Disposition: Possible (ability to occur), Necessary (negation is not possible);

C2) Subjective (related to the observer)

C2a) Believable (observer accepts data around 50%, but is ready to reconsider it),

C2b) Probable (observer accepts data around 80%; however, is not ready to reconsider it),

C2c) Doubtful (observer can hardly accept data),

C2d) Possible (observer considers that data could be true),

C2e) Irrelevant (observer does not care about the data),

C2f) Undecidable (observer is not able to decide if true or false).

Thus, inconsistency appears when there is no world ω that is compatible with the evidence e. Usually the consistency of information is not graduated; we speak only of consistent or inconsistent information.

We may consider that the evidence e induces a partial order among the worlds: some worlds ω_1 are more compatible with evidence e than other worlds ω_2. Usually this partial order is connected with the "quantity of changes" that a world has to undertake so that it becomes compatible with evidence e.)

If several worlds $\omega_1, \omega_2, \dots$ are compatible with evidence e, we say that we have an imprecision. Thus, evidence e is precise if only one world is compatible with e.

In what follows we will be interested in probabilistic reasoning. As an example, consider the following: around the logical reasoning (supposed perfect)

"If a patient has a toothache and there is a large black hole in the aching tooth, then the patient has a cavity"

we have at least these two if-then rules, apparently of the same kind:

"If a patient has a toothache, then there is a probability of 0.9 that he/she has a cavity"

"If the patient has a cavity, then there is a probability of 0.4 that he/she has a toothache".

What the numbers 0.9 and 0.4 mean and how they are related? A possible answer is given by the classical Probability Theory.

3.2 Usual and Precise Meaning

We live in a continuously changing world, we receive incomplete information and we possess imperfect knowledge about it. Still, we have to take decisions that affect our state. Of course, we would like to foresee the effects each possible decision could cause and we are endowed, naturally, with an evaluation system. However, in general the effects are "not sure", each of them is more or less likely to be observed. Our past experiences, combined with our reasoning system, help us to evaluate the respective likelihood, which we call also chance.

Our everyday life is governed by a number of "happenings" and their large number usually prevents us to fully understand what "happens in reality". A first step toward the comprehension of life consists in studying isolate simple deterministic cause-effect phenomena, such as:

– If a stone (coin, dice) is released from our hand, then it falls down to earth
– If the temperature increases over 0°C, then the ice melts
– The iron will rust because of the existence of oxygen in the air

and so on. In some cases formulas – even elaborate and complex – are found, which allow us to compute in advance a future situation. However, our imperfect knowledge prevents us to make fully reliable predictions and forces us to speak about "chances", "probabilities", and "random behavior".

The meaning of the words *probability* and *random* seems obvious for everybody. Usually we "understand correctly" these words in sentences expressed in a natural language. However, our intuition is often misleading.

Let us consider some examples, taken from examination papers (in reputed English universities [Lindley 1965]):

1) Circular discs of radius r are thrown onto a circular plane table of radius R that is surrounded by a border of uniform width r lying in the same plane as the table. If the discs are thrown independently and at random, and N discs stay on the table, show that the probability that a fixed point on the table will be covered is …

2) Color blindness appears in 1% of the people in a certain population. How large must a random sample (with replacement) be if the probability of its containing a color blind person is 0.9?

3) In each individual there is a pair of genes, each of which may be of type **X** or of type **x**. An individual with **xx** is considered abnormal. Each of the parents of a child transmits one of its own genes to the child. Mating in the population is to be assumed to be random. Show that among normal children of normal parents the expected proportion of abnormal is …

4) A patient with a needle 5 cm long in his chest is X-rayed. If the orientation of the needle is quite random, what is the distribution of the length of the needle's shadow?

5) A radioactive source emitting alpha particles in random directions (all equally likely) is held at distance d from a plane photographic plate (supposed infinite). What is the distribution of …?

Another example, taken recently from the Internet:

6) In the UK 50.9% of all babies born are girls; suppose then that we are interested in the event A: "a randomly selected baby is a girl". According to the frequentist approach its probability is 0.509.

We feel that probability is somehow expressing chances of appearance of an event. However, "random" is more difficult to be explained; in our minds it should have to do something with equally likely and perhaps the best illustration of the word is given by assuming the use of a (perfect) dice.

There are other ways to express chances of appearances of an event; for example in bookmaker's idiom the word odds is doing the job.

Careless use of words may easily lead to paradoxes, and the use of probability and random without proper definitions makes no exception. This is one reason why we need precise definitions for these terms (concepts).

People interested in chance games and insurance policies initiated long ago the study of probabilities. The Chevalier de Méré is remembered as to ask Pascal[1] to compute gain chances in dice games; the solution was given in 1654 in a letter to Pierre Fermat. More results by Huygens[2] appeared in 1657 in *De Ratiociniis in Ludo Aleae* and were presented also by Jacques Bernoulli[3] in his book *Ars Conjectandi*, which was published posthumous in 1713. In *De Mensura Sortis* and in *Doctrine of Chances*, which appeared in 1711 res. 1718, de Moivre initiated insurance calculus.

A name that should appear in any book of probabilities is Thomas Bayes[4]. In a famous paper *An essay towards solving a problem in the doctrine of chances*, published posthumous (1763), some basic ideas about how to treat probabilities were presented.

Many paradoxes, due to loose axiomatic approach when working with probabilities, lead during the 19th Century to serious developments in mathematical analysis. During the 20th Century the accent moved to statistics, despite the incomplete foundation of probability theory. The first axiomatic treatment of probabilities appeared in the treatise of A. N. Kolmogorov published in 1933.

Still today there are different opinions about how to interpret the word "probability". Perhaps it is best to emphasize the ideas of Bruno de Finetti: "the probability (of an event) is an expression of the agent's view of the world and as such it has no existence of its own".

Usual meaning of "probability" is linked to a particular "event". In what follows we present the classical approach to probabilities, which is based on a "calculus" with events.

3.3 Experiments and Events

In mechanics, when studying the evolution of a mechanical system, we encounter a lot of laws having a deterministic character, as for example the law of falling bodies. This law allows *You* to very accurately predict the trajectory of a bullet or a rocket. We are convinced that if the "experience" is repeated, exactly the same result will be obtained.

On the contrary, in thermodynamics most of the laws are not of deterministic type. The thermodynamic phenomena are random; it is impossible to predict a certain result of such a phenomenon or process; however, *You* may predict a distribution of results.

---

[1] Blaise Pascal (1623-1662), Pierre Fermat (1601-1665), Abraham de Moivre (1667-1754), French mathematicians.

[2] Christiaan Huygens (1629-1695), Dutch mathematician and astronomer.

[3] Jacques Bernoulli (1654-1705), Swiss mathematician.

[4] Thomas Bayes (1702-1761), British Presbyterian minister.

The biological laws (starting with Mendelian inheritance laws) have all a random character.

By **experiment** we understand a repeatable process, having an identifiable or a measurable outcome. Typical examples are:

- After throwing a dice we obtain an identifiable outcome;
- After measuring (in microns) the diameter of a cell, or the weight of a person (in grams), or the systolic pressure of a man, we obtain measurable outcomes;
- After looking at a pulmonary X-ray of a patient we obtain an identifiable outcome;
- After counting the flu cases detected by a physician we obtain an identifiable outcome (even if this is expressed as a number).

(Recall that measuring an object means comparing it with another object, previously chosen as a measure unit, and not counting!)

In most cases – and usually in biology – the outcome of an experiment is not unique and cannot be accurately predicted; we say that the experiment is **random**. This means in fact that if several times a random experiment is repeated, each time a different result could be obtained.

From the logical point of view, the possible outcomes cannot be decomposed. Accordingly, by an **elementary event** we understand a possible outcome of a random experiment. An **event** is simply a collection of elementary events.

Obvious examples are obtained when throwing a dice (with six possible outcomes). Many ordinary people consider only the appearance of ⚅ as an "event". However, according to our definition above, the appearance of ⚁ is an elementary event and an event is also the appearance of ⚁ or ⚃! After such an experiment, a lot of events – not only six, but also fifty-eight others – may appear!

As another example, an "event" is the detection of a tumor when regarding the pulmonary X-ray of our patient "John Johnson". (Here we encounter a simpler situation, only two outcomes are possible: we detect, or we do not detect the tumor. However, we will see later that we have here, according to the theory, four different events!)

Consider for example the experiment consisting of measuring the diameter of a cell, which gives us a measurable result. If we choose an arbitrary interval $[a, b]$ of real numbers – where $a < b$ – we obtain the following event: the result of measuring a cell's diameter falls into our interval, that means between a and b. Denote by E this particular event.

From a logical point of view, we may consider also the complementary event: the result of measuring the cell's diameter falls outside our interval, that means either before a, or after b. This **complement** of E will be denoted $\overline{E}$ (read "E bar").

Of course, if our interval is [0, 1 (km)], then it is "certain" that the result of measuring cell diameters will fall into this interval. In this case we obtain the **certain event**. The complementary of the certain event is called the **impossible event** and is denoted by the symbol (the same used in set theory to denote the void set).

(When throwing a dice, the certain event consists of the appearance of any face; the impossible event consists of not appearing any face at all.) Let us continue now to measure the diameter of cells and let us choose another interval $[c, d]$ of real numbers ($c < d$). Denote by F the following event: the result of measuring a diameter falls in this latter interval, namely between c and d.

Now we have two genuine events, E and F (and automatically two complementary events, $\overline{E}$ and $\overline{F}$). However, logic tells us that another event appears: the result of measuring a diameter falls between c and b (see Figure 3.1 below).

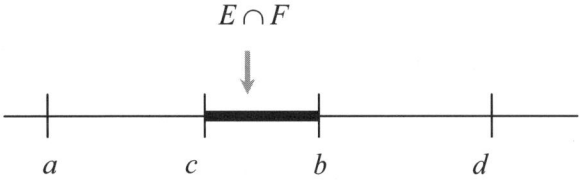

Fig. 3.1. Conjunction of events: example

It is natural to denote this latter event by $E \cap F$ and to say that it is the **conjunction** of events E and F. We will read this as "the event E and F".

In general, given the events E and F as outcomes of the same experiment, we can imagine another event $E \cap F$ as an outcome of our experiment.

For example, if we throw a dice, and if E denotes the appearance of a "less than three points" face, and F denotes the appearance of an "even" face, then $E \cap F$ is exactly the appearance of [die face showing 2].

However, if E is the appearance of [die face showing 4] and F is the appearance of [die face showing 6], then obviously $E \cap F$ is impossible, i.e.
$$E \cap F = \varnothing.$$

In general, if for two events E and F one has
$$E \cap F = \varnothing,$$
it is said that the events **are exclusive**.

From a logical point of view, given two events E and F, we could consider a disjunction $E \cup F$. In our "measuring cells" example above, this event could be interpreted as follows: the result of measuring a cell's diameter falls between a and d (see Figure 3.2 below).

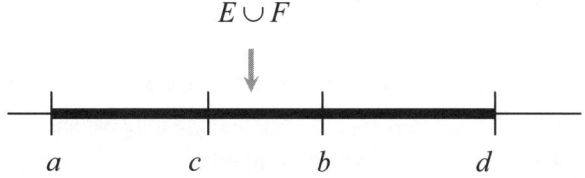

Fig. 3.2. Disjunction of events: example

Attention, we may have other situations, such as that illustrated in the Figure 3.3 below; here $E \cup F$ means that the results of measuring diameters falls between a and b, or between c and d.

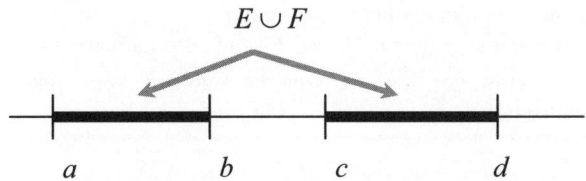

Fig. 3.3. Disjunction of events: another example

The three operations with events, presented above (complement, conjunction, and disjunction) allow us to construct a calculus with events. We do not insist on this calculus, because it is analogous to the calculus with sets. The notations used suggest this also.

This analogy is the basis of representing events as subsets of a "universe" Ω; then the conjunction of events becomes the intersection of the corresponding subsets (see Figure 3.4), and the disjunction of events becomes the union of the corresponding subsets. Of course, Ω itself represents the certain event.

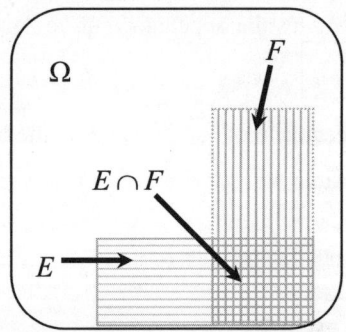

Fig. 3.4. Conjunction of events in the universe

Consider now a finite family $E_1, E_2,..., E_n$ of events. If the disjunction of these events coincides with the certain event, then we say that the family **is exhaustive**.

If each two events E_i, E_j $(i \neq j)$ from the family are exclusive, i.e. $E_i \cap E_j = \varnothing$, we say that the events in the family **are mutually exclusive**.

The most interesting situation is that of a mutually exclusive and exhaustive family of events. In this case the family is called **complete**.

As a very simple example, consider a dice thrown on the table. Denote E_1 the appearance of the face ⚅, E_2 the appearance of ⚁ or ⚁, E_3 the appearance of any other face. It is obvious that the family E_1, E_2, E_3 is complete.

The use of notations from Set Theory is not casual. Let us explain why.

3.4 Formal Definition of Events

Consider a "universe" Ω of "possible worlds". To be formally correct, let us define an **algebra** (over the universe Ω) as a family $\mathcal{A}$ of subsets of Ω that satisfies two conditions:

(S1) $\Omega \in \mathcal{A}$,

(S2) $X - Y \in \mathcal{A}$ for every $X, Y \in \mathcal{A}$.

(Recall the difference $X - Y$ is the set of elements of Ω that belong to X, not also to Y. The complement of subset X is exactly $\Omega - X$.)

Proposition 3.1. If $\mathcal{A}$ is an algebra (over the "universe" Ω), then it satisfies:

(I) $X \cap Y \in \mathcal{A}$ for every $X, Y \in \mathcal{A}$,

(U) $X \cup Y \in \mathcal{A}$ for every $X, Y \in \mathcal{A}$.

Indeed, condition (I) follows from the equality $X \cap Y = X - (\Omega - Y)$ that expresses intersection in terms of the difference. As for (U), the equality $X \cup Y = \Omega - ((\Omega - X) - Y)$ is used; hence the union can also be expressed in terms of the difference.

A **sigma-algebra** (over Ω) is a family S of subsets of Ω that satisfies the following three conditions:

(S1) $\Omega \in S$,

(S2) $X - Y \in S$ for every $X, Y \in S$,

(S3) If $X_1, X_2, ..., X_n, ...$ is a countable family of subsets from S, then

$$\bigcup_{n=1}^{\infty} X_n \in S.$$

Recall the countable union $\bigcup_{n=1}^{\infty} X_n$ of the family $\{X_1, X_2, ..., X_n, ...\}$ is the set of all elements of Ω that belong to at least a member X_n of the family.

The countable intersection $\bigcap_{n=1}^{\infty} X_n$ of a countable family of subsets from the sigma-algebra S belongs to S, because of the relation

$$\bigcap_{n=1}^{\infty} X_n = \Omega - \bigcup_{n=1}^{\infty} (\Omega - X_n).$$

The superior limit of the family above, denoted $\limsup_{n} X_n$, is the set of all elements of Ω that belong to X_n for infinitely many n. Thus,

$$\limsup_{n} X_n = \bigcap_{n=1}^{\infty} \bigcup_{k=n}^{\infty} X_k$$

and it is obvious that $\limsup_{n} X_n \in S$ when all $X_n \in S$.

Dually, the inferior limit of the family $\{X_1, X_2, ..., X_n, ...\}$, denoted $\liminf_{n} X_n$, is the set of all elements of Ω that belong to X_n for all except a finite number of n.

Thus, $\liminf_{n} X_n = \bigcup_{n=1}^{\infty} \bigcap_{k=n}^{\infty} X_k$.

The inclusion $\liminf_{n} X_n \subseteq \limsup_{n} X_n$ is always true. When the equality $\liminf_{n} X_n = \limsup_{n} X_n$ takes place, the common value is called the limit of the family $\{X_1, X_2, ..., X_n, ...\}$.

Hence, a sigma-algebra contains, together with a countable family of subsets, its superior limit and its inferior limit.

Examples. 1) Of course, if Ω is an arbitrary set, then its power set 2^{Ω} is a sigma-algebra.

Recall the elements of the power set 2^{Ω} are the functions $\chi : \Omega \rightarrow \{0, 1\}$.

2) Consider $\Omega = \mathbf{N}$ the set of natural numbers and, for a subset X, define $X \in \mathcal{A}$ if either X is finite, or the complement $\mathbf{N} - X$ is finite. Then $\mathcal{A}$ is an algebra, but not a sigma-algebra.

Indeed, the subset of even natural numbers is not in $\mathcal{A}$. However, this is a countable union of finite subsets!

3) Consider $\Omega = \mathbf{R}$ the set of real numbers. Start with the family $\mathcal{J}$ of left closed, right open intervals

$$[a, b) = \{x \mid x \in \mathbf{R} \text{ and } -\infty < a \leq x < b < \infty\}$$

where $a < b$ are real numbers.

Define the family $\mathcal{D}$ of countable unions of intervals, i.e. $D \in \mathcal{D}$ iff there exists a finite family $\{J_1, J_2, ..., J_n\}$ where each $J_k \in \mathcal{J}$, such that $J_k \cap J_l = \varnothing$ for $k \neq l$, and $D = \bigcup_{k=1}^{n} J_k$.

Now, define the family $\mathcal{S}$ as follows: $S \in \mathcal{S}$ iff either $S \in \mathcal{D}$, or $\mathbf{R} - S \in \mathcal{D}$.

Finally, define the family $\mathcal{B}$ as the "smallest" sigma-algebra containing $\mathcal{S}$. $\mathcal{B}$ is known as the Borel[5] sigma-algebra, and its elements are known as the Borel sets in $\mathbf{R}$.

If we repeat the construction above for $\Omega = \mathbf{R}^m$, taking as intervals in $\mathcal{J}$ the Cartesian products (see Figure 3.5 for an example)

$$\underset{i=1}{\overset{m}{\times}} [a_i, b_i),$$

then we obtain the m-dimensional Borel sigma-algebra.

Fig. 3.5. 2-dimensional Borel set

Now, consider a sigma-algebra $\mathcal{E}$ (over Ω). Its elements will be called **events**. In this case Ω is called the **event space**.

Consider the set $\Omega_1 = \{H, T\}$ related to a fair coin toss. The sigma-algebra 2^{Ω_1} describes the possible results that may appear after one toss. On the other hand, the possible results of the examination of patients' pulmonary X-ray are described by a similar sigma-algebra 2^{Ω_2} where $\Omega_2 = \{\text{tumor detected, tumor not detected}\}$. Despite the similarity of the sigma-algebras, we feel there is a difference, and what make the difference are the different chances.

3.5 Defining Probabilities

Chances of events can be modeled by probabilities. As was pointed out above, the systematic study of probability has started around 1650. Since then at least three main

---

[5] Émile Borel (1871-1956), French mathematician.

approaches to define probability have been accepted: logical, frequentist, and subjective (Bayesian).

In the logical (classical) approach, we identify elementary outcomes in a way that makes them equally possible (using, for example, the assumptions of symmetry and homogeneity). Then, given an event E, we attach its probability $P_L(E)$ by dividing the number of elementary outcomes favorable to E to the total number of elementary outcomes.

As an example, picking a club card from an ordinary deck of cards is an event related to 13 elementary outcomes from a total of 52, so its probability is $\frac{13}{52} = 0.25$. (However, how we do know that the choice of the card is "random", not related to any prior knowledge about the arrangement of cards? We do not.)

As another obvious example, the probability of a head appearing after tossing of a fair coin is $\frac{1}{2} = 0.5$. (However, how do we know that the coin is "fair", i.e. symmetric and homogeneous? We do not.)

In the frequentist approach, we restrict attention to phenomena that are inherently repeatable under identical conditions. Then we define the probability $P_F(E)$ as the limiting value of relative frequency with which the event E occurs when the number of repetitions tends to infinity. Of course, to estimate $P_F(E)$ we limit ourselves to a "big" number of repetitions.

As an example of frequentist approach, let us estimate the probability of a rain falling over Toronto tomorrow. Consider data registered in 200 days and count the number of rainy days (let us say 84); assume that each day is "identical" to each other (i.e. has the same chances to be a rainy day), so the relative frequency is $\frac{84}{200} = 0.42$; we estimate our probability by its frequency.

Thomas Bayes highlighted the main critics of the above definitions and his ideas are best explained when considering tossing a coin. The number $\frac{1}{2} = 0.5$ accepted as the probability of a head appearing after a (single) toss is neither a property of the coin, nor of the experimental setup that generates the outcome. This is an opinion of the agent *You* who observes the experience and, due to its limited knowledge, cannot predict exactly the outcome! When our agent *You* must decide something that depends on the outcome of the toss, his estimate of the probability is obtained as follows – he would pay an amount to enter in a game, in which he would gain 1 when the head appears, res. 0 in the contrary case. The maximal amount that the agent *You* is willing to pay is the probability of the event.

More precisely, in the Bayesian approach, we imagine a bet about the truth of a proposition e = "the event E will appear", against an opponent (which considers as true the negation $\neg e$); we evaluate the odds $O(e)$ in favor of e against $\neg e$ when the bet is judged "fair"; from this value the probability of the event E will be, by definition, $P_B(E) = O(e) \Big/ (1 + O(e))$.

Of course, any other person is free to evaluate his odds in favor of e against $\neg e$ at a value different from our $O(e)$! Moreover, any change in our state of knowledge may lead us to re-evaluate the odds!

What kind of probability is appropriate, and how *You* would assess $P(E)$ – i.e. the probability of the event E – in each case?

Each of the three approaches above has some advantages and some weaknesses (drawbacks). Indeed, the logical approach is very simple when applied to (ideal) coin tossing, (ideal) dice throwing, drawing (ideal) balls from urns a.s.o. However, what means "equi-possible"? We have no grounds – mainly because of insufficient information – for favoring an elementary outcome over another; that is we implicitly use the **principle of insufficient reason**. And this is leading to paradoxes!

The probability $P(E)$ refers always to a future possible event E and is estimated now. It has no meaning if we know that the event "has appeared".

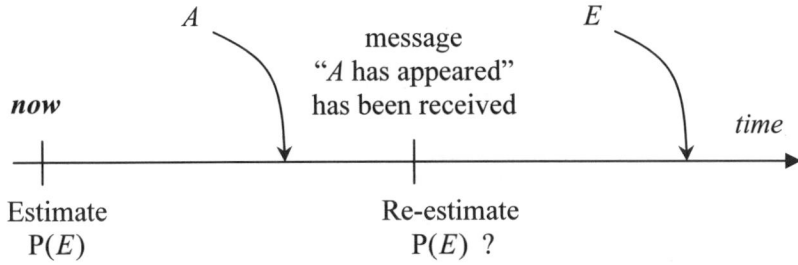

Fig. 3.6. Re-estimating probabilities

In Figure 3.6 a typical situation is presented. We estimate *now* the probability of a future event E. However, as times goes by, we receive a message stating that "event A has appeared" and we know that this could modify the chances of E. We are obliged to re-estimate the probability of (still future) E.

If we are *now* and we think in advance of such a situation, then the re-estimate should be denoted differently from $P(E)$. A classical notation is $p(E\,|\,A)$, read as "the probability of E conditioned by A".

A good example of conditioning events is taken from actuaries. Consider the following sentence:

"The probability that a man aged 40 will die within 10 years from now is 15%".

Here the event E is death within 10 years from now and the event A is "the man will reach age 40".

Let us give a more relevant example, from medicine. Suppose we have patients suffering or not from a disease D, and we imagine a test S, which could give either positive, or negative results. Sometimes we use the sentence:

"The patient tests positive provided he/she has disease D"

and more often we use the sentence:

"The patient has disease D, provided he/she tested positive".

Both are examples of conditional events. It is essential which event comes first.

3.6 Defining Probabilities (II)

To formally define probabilities, let us consider a universe Ω and a sigma-algebra $\mathcal{E}$ (over Ω) whose elements are the events.

We will interpret $P(E)$ as the area of the subset E, accepting that the universe Ω has a unit area (see Figure 3.7).

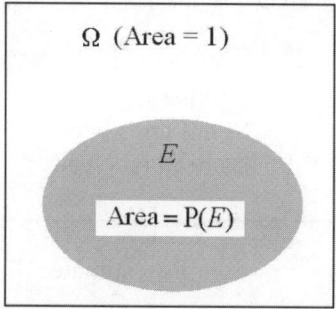

Fig. 3.7. Probabilities – geometrical representation

A first and immediate relation is the so-called **relation of the complement**: if E is an event and we "know" its probability, then we "know" also the probability of the complement

$$P(\overline{E}) = 1 - P(E).$$

As a consequence, if E has a "big" probability, then the complement $\overline{E}$ has a "small" probability. In particular, the impossible event has null probability $P(\varnothing) = 0$.

The **addition relation** is easy to express: if E and F are two arbitrary events, then

$$P(E \cup F) = P(E) + P(F) - P(E \cap F)$$

i.e. "knowing" probabilities $P(E)$ and $P(F)$, we "know" also $P(E \cup F)$, provided we "know" $P(E \cap F)$! This relation is immediate if we look at Figure 3.4: if we add the areas of E and F we obtain the area of $E \cup F$, but we notice that the area of the intersection $E \cap F$ was counted twice!

The *a priori* realization of the event A restrains the universe (from Ω to the subset A – see Figure 3.8). $E \mid A$ is represented by the intersection of the two subsets, but if we want to calculate its probability, we have to report ourselves to the new universe A instead of Ω (and the area of A should be 1 in this new situation). Therefore,

$$p(E \mid A) = \frac{P(E \cap A)}{P(A)}$$

(obviously, valid when $P(A) \neq 0$).

Two events E and F are called **independent** if no one is conditioning the other, which means the *a priori* realization of one is not changing the probability of the other:

$$p(E \mid F) = P(E) \text{ and } p(F \mid E) = P(F) .$$

For example, if we have two coins and we throw them separately, then it is accepted that the appearances of the head in each one are independent events.

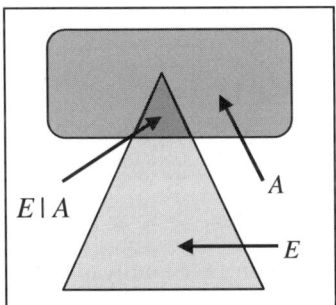

Fig. 3.8. Geometrical representation of conditional events

The **relation of the independent events** is the following:

$$P(E \cap F) = P(E) \cdot P(F) .$$

This relation allows us to say that the probability of obtaining a double six (when throwing two independent dice) is $\frac{1}{36}$.

Let us give now a formal definition of probabilities. As above $\mathcal{E}$ is the sigma-algebra of events (over the universe Ω). The starting point is the definition of conditional probabilities.

A **probability** is simply a (two-argument) function

$$p : \mathcal{E} \times \mathcal{E} \rightarrow [0,1] ,$$

which satisfies the following conditions:

(P1) $p(A \mid A) = 1$ for all $A \in \mathcal{E}$.

(P2)   Finite additivity

If $A_1, A_2 \in \mathcal{E}$ and $B \in \mathcal{E}$, and if $A_1 \cap A_2 \cap B = \varnothing$, then

$$p(A_1 \cup A_2 \mid B) = p(A_1 \mid B) + p(A_2 \mid B).$$

(P2') Countable additivity

If $B \in \mathcal{E}$ and $A_1, A_2,...$ is a countable family of events such that $(A_i \cap B) \cap (A_j \cap B) = \varnothing$ for $i \neq j$, then

$$p(\bigcup_{n=1}^{\infty} A_n \mid B) = \sum_{n=1}^{\infty} p(A_n \mid B).$$

(P3) $p(A \cap B \mid C) = p(A \mid C) \cdot p(B \mid A \cap C)$ for all $A, B, C \in \mathcal{E}$.

Of course, we say that $p(A \mid B)$ is the probability of the event A conditioned by B, or the conditional probability of the event A given B.

From the above axioms, we deduce the following important results:

Proposition 3.2. $p(\Omega \mid A) = 1$ for all $A \in \mathcal{E}$.

Indeed in (P3) let us put $B = C = \Omega$. Then,

$$p(A \cap \Omega \mid \Omega) = p(A \mid \Omega) \cdot p(\Omega \mid A \cap \Omega) \text{ or } p(A \mid \Omega) = p(A \mid \Omega) \cdot p(\Omega \mid A).$$

Proposition 3.3. $p(A \mid B) + p(\Omega - A \mid B) = 1$ for all $A, B \in \mathcal{E}$.

Indeed, in (P2) let us put $A_1 = A$, $A_2 = \Omega - A$. Then,

$$p(\Omega \mid B) = p(A \mid B) + p(\Omega - A \mid B).$$

In general, if $A_1, A_2,..., A_n$ is a complete family of events then

$$\sum_{i=1}^{n} p(A_i \mid B) = 1 \text{ for all } B \in \mathcal{E}.$$

Remark. Starting with the function p, we obtain by

$$P(E) = p(E \mid \Omega)$$

the "probability" of A that we used before. This function

$$P : \mathcal{E} \rightarrow [0,1]$$

satisfies the conditions:

(Π1) $P(\Omega) = 1$,

(Π2) $P(\bigcup_{n=1}^{\infty} A_n) = \sum_{n=1}^{\infty} P(A_n)$ if $A_i \cap A_j = \varnothing$ for $i \neq j$,

and in particular

$$P(A_1 \cup A_2) = P(A_1) + P(A_2) \text{ if } A_1 \cap A_2 = \emptyset.$$

From (P3) we obtain the formula for the conditional probability:

$$P(A \cap B) = P(A) \cdot p(B \mid A).$$

Examples. 1) Toss a die. Then $\Omega = \{1, 2, 3, 4, 5, 6\}$. Here we consider that $\mathcal{E}$ is the power set (i.e. the set of all subsets) of Ω and has 64 elements.

If $A = \{2, 6\}$ and $B = \{4, 5, 6\}$, then the event A conditioned by B is $A \mid B = \{6\}$.

Also $A \mid (\Omega - B) = \{2, 4\}$ and $B \mid (\Omega - A) = \{4, 5\}$. The values $P(A) = \dfrac{1}{3}$, $P(B) = \dfrac{1}{2}$, $p(A \mid B) = \dfrac{1}{3}$ are immediate.

2) Consider the real numbers $\mathbf{R}$, and we want to define "probability".

We have to choose first the set $\mathcal{E}$ of "events" of $\mathbf{R}$. We choose this set such that it contains all left-closed, right open intervals.

Because $\mathcal{E}$ is a sigma-algebra, this set contains all countable unions of intervals. Each set A of $\mathcal{E}$ can be expressed as a countable union of intervals.

How can we define the functions P and p? There are several possibilities. Here is one:

Consider a continuous function $\phi : \mathbf{R} \to \mathbf{R}$ such that all its values are strictly positive, and also $\displaystyle\int_{-\infty}^{\infty} \phi(x)\, dx = 1$.

If $A = [a, a')$ and $B = [b, b')$ are intervals such that $a < b < a' < b'$, then

$$P(A) = \int_a^{a'} \phi(x)\, dx \text{ and } p(A \mid B) = \int_b^{a'} \phi(x)\, dx \bigg/ \int_b^{b'} \phi(x)\, dx.$$

3.7 Bayes' Theorem

The first result of Thomas Bayes is easy to state and to prove. It states that if A, B are events and we know the probabilities $p(A \mid B)$, $P(A)$ and $P(B)$, then $p(B \mid A)$ is computed as follows:

$$p(B \mid A) = \frac{p(A \mid B) \cdot P(B)}{P(A)}.$$

Indeed, we use twice the relation of conditional probability

$$P(A \cap B) = P(A) \cdot p(B \mid A), \; P(B \cap A) = P(B) \cdot p(A \mid B)$$

and we exploit the commutativity of the conjunction of events.

As an immediate application of this result, we obtain a very important probability

$$p(disease \mid test) = \frac{p(test \mid disease) \cdot P(disease)}{P(test)}$$

provided all three probabilities in the right hand of the formula have good estimations.

As for the second result of Bayes, it has the following mathematical expression:

Theorem 3.1. If $D_1, D_2,..., D_n$ is a complete family of events, whose *a priori* probabilities $P(D_1), P(D_2),..., P(D_n)$ are known, and if S is another event such that all conditional probabilities

$$p(S \mid D_1), p(S \mid D_2),..., p(S \mid D_n)$$

are known, then the reversed conditional probabilities $p(D_i \mid S)$ are obtained from the formula:

$$p(D_i \mid S) = \frac{P(D_i) \cdot p(S \mid D_i)}{P(D_1) \cdot p(S \mid D_1) + P(D_2) \cdot p(S \mid D_2) + ... + P(D_n) \cdot p(S \mid D_n)}.$$

Proof. Recall that if $D_1, D_2,..., D_n$ form a complete family of events, then $D_1 \cup D_2 \cup ... \cup D_n = \Omega$ and $D_i \cap D_j = \varnothing$ for $i \neq j$.

To prove the theorem it is enough to establish that

$$P(S) = P(D_1) \cdot p(S \mid D_1) + P(D_2) \cdot p(S \mid D_2) + ... + P(D_n) \cdot p(S \mid D_n).$$

Denoting $A_i = S \cap D_i$, we have $S = \bigcup_{i=1}^{n} A_i$ and $A_i \cap A_j = \varnothing$, hence $P(S) = \sum_{i=1}^{n} P(A_i)$ from condition (Π2). Now $P(A_i) = P(D_i) \cdot p(S \mid D_i)$ and the theorem is proved.

Bayes' theorem is a tool used to re-evaluate the probabilities of different diagnostic hypotheses.

The notations $D_1, D_2,..., D_n$ above may refer to these diagnostic hypotheses. The *a priori* probabilities $P(D_i)$ are estimated by different methods, for example by using national statistical surveys. The event S is a sign or symptom. In the context of the presence of this sign, the probabilities of diagnostic hypotheses should be reevaluated; the Bayes' theorem above provides us with a formula to calculate these *a posteriori* probabilities.

Let us consider the following simple example:

D_1 – our patient has TB,

$D_2 = \overline{D_1}$ – our patient has not TB,

S – patient "John Johnson" tests positive (in the pulmonary X-ray).

Of course, D_1, D_2 is assimilated to a complete family of events. From national statistical data we know that 3% of the population has TB. Therefore we estimate:

$$P(D_1) = 3\% = 0.03 \ ;$$

consequently

$$P(D_2) = 1 - 0.03 = 0.97 \ .$$

We estimate now $p(S \mid D_1)$. From medical experience we know that 90% of the patients with TB test positive in the pulmonary X-ray. Hence, $p(S \mid D_1) = 0.90$. Also, there is a small chance, let us say 1%, that a non-TB person tests positive. Hence $p(S \mid D_2) = 0.01$.

We have now all the ingredients for the Bayes' formula:

$$p(D_1 \mid S) = \frac{0.03 \times 0.90}{0.03 \cdot 0.90 + 0.97 \cdot 0.01} = \frac{0.027}{0.0367} = 0.736$$

Therefore, the probability that "John Johnson", who tested positive in the pulmonary X-ray has TB is estimated at 73.6%.

(That is, because of the event S, the probability of a tb diagnosis raised from 3% to 73.6%.)

3.8 Misleading Aspects

It seems obvious that, given an event E, some events may favor and some others disfavor our event. Inexperienced people could "believe" that, if an event A favors another event B and the event B favors a third event C, then A should favor also C.

Let us try to define what "favor" means. It seems reasonable to accept that $p(B \mid A) > P(B)$, i.e. the appearance of A will increase the probability of B, should be a good definition for "A favors B". Hence, the transitivity reasoning (similar to a syllogism)

$$\frac{A \quad \text{favors} \quad B, \quad B \quad \text{favors} \quad C}{A \quad \text{favors} \quad C}$$

is replaced by a simple implication

$$\text{if} \quad \begin{matrix} p(B \mid A) > P(B) \\ p(C \mid B) > P(C) \end{matrix} , \text{then} \ p(C \mid A) > P(C) .$$

However, it is not difficult to find a counter-example. Consider the event space

$$\Omega = \{00000, 00100, 01100, 10000, 11000, 11001, 11010, 11100\}$$

whose "worlds" ω are described binary, and denote by A_k the event "the kth digit of ω is 1}. It is easy to establish that $P(A_1) = P(A_2) = \dfrac{5}{8}$, $P(A_3) = \dfrac{3}{8}$ and that $p(A_2 \mid A_1) = \dfrac{4}{5} > \dfrac{5}{8}$, $p(A_3 \mid A_2) = \dfrac{2}{5} > \dfrac{3}{8}$, i.e. A_1 favors A_2 and A_2 favors A_3. However, $p(A_3 \mid A_1) = \dfrac{1}{5} < \dfrac{3}{8}$, hence A_1 does not favor A_3!

3.9 Random Variables and Distributions

"Random variables" and "distributions" provide another way to describe results of some particular random experiments.

In general, a variable may take several values. Univariate variables are supposed to take values from $\mathbf{R}$, the set of real numbers.

A set D of values is called discrete if for each distinct "points" $v, w \in D$ there exist open intervals $I_v = (a,b)$, $I_w = (c,d)$ such that $v \in I_v$, $w \in I_w$ and $I_v \cap I_w = \varnothing$.

Discrete subsets of $\mathbf{R}$ are necessarily finite or countable, thus their values could be "listed". (However, in general, countable subsets of $\mathbf{R}$ are not necessarily discrete!)

The expression "X is a **discrete random variable**" means:

- The possible values of the variable $x_1, x_2, \ldots, x_n, \ldots$ are numeric and discrete, and
- We know all the probabilities $p_1, p_2, \ldots, p_n, \ldots$ associated to these values.

The values $x_1, x_2, \ldots, x_n, \ldots$ form the so-called **domain** of the random variable X.

A discrete random variable X is perfectly described by its **distribution table**:

$$\begin{pmatrix} x_1 & x_2 & \ldots & x_n & \ldots \\ p_1 & p_2 & \ldots & p_n & \ldots \end{pmatrix}.$$

where in the first row the elements of the domain of X are listed. The probabilities on the second row are real numbers that satisfy two conditions:

1) All are positive, i.e. $p_n \geq 0$ for each value x_n.
2) Their sum is 1:

$$p_1 + p_2 + \ldots + p_n + \ldots = 1 .$$

The number p_n is interpreted as the "chance" that the random variable X will take, in a future experiment, exactly the value x_n. In other words, p_n is the probability of the event $X = x_n$, i.e.

$$p_n = P(X = x_n) .$$

(Here the event space Ω is exactly the domain of the random variable. The events are all the subsets of Ω.)

The distribution table shows how the "chances" are distributed among the possible values of the random variable. For this reason we use the word "distribution" instead of the longer "random variable".

When the values are ordered:

$$x_1 < x_2 < \ldots < x_n < \ldots$$

cumulative probabilities are easily obtained:

$$P(X \leq x_n) = \sum_{i=1}^{n} p_i .$$

Conversely, if the cumulative probabilities are known, then the ordinary probabilities p_n for $n > 1$ are obtained as follows:

$$p_n = P(X \leq x_n) - P(X \leq x_{n-1}) .$$

As an example, the following table

$$\begin{pmatrix} \mathbf{AA} & \mathbf{Aa} & \mathbf{aa} \\ \dfrac{9}{16} & \dfrac{6}{16} & \dfrac{1}{16} \end{pmatrix}$$

may describe the genotype (of a specific gene, with alleles **A** and **a**). Obviously, $\dfrac{9}{16}$ is the probability that an individual of the population, chosen at random, belongs to genotype **AA**.

Expressing the genotype **AA** by number 0, the genotype **Aa** by 1 and the genotype **aa** by 2, the table

$$\begin{pmatrix} 0 & 1 & 2 \\ \dfrac{9}{16} & \dfrac{6}{16} & \dfrac{1}{16} \end{pmatrix}$$

describes a distribution of probability. The corresponding random variable could be interpreted as "the number of dominated alleles in the genotype".

3.10 Expectation and Variance

In general, given a discrete random variable:

$$X : \begin{pmatrix} x_1 & x_2 & \ldots & x_n & \ldots \\ p_1 & p_2 & \ldots & p_n & \ldots \end{pmatrix}$$

if we can compute the value

$$E(X) = x_1 \cdot p_1 + x_2 \cdot p_2 + ... + x_n \cdot p_n + ...$$

this value is known as the **expectation** of the variable X. It is in fact the "weighted average" of the possible values $x_1, x_2, ..., x_n, ...$, the weights being exactly the probabilities $p_1, p_2, ..., p_n, ...$. The number $E(X)$ could be thought of as a "center" of the values of X.

For example, for the discrete random variable

$$\begin{pmatrix} 2 & 4 & ... & 2^n & ... \\ \frac{1}{2} & \frac{1}{4} & ... & \frac{1}{2^n} & ... \end{pmatrix}$$

the value

$$2 \cdot \frac{1}{2} + 4 \cdot \frac{1}{4} + ... + 2^n \cdot \frac{1}{2^n} + ...$$

is not a number, thus "$E(X)$ cannot be computed".

Given a discrete random variable X and a real number r, the table

$$X - r : \begin{pmatrix} x_1 - r & x_2 - r & ... & x_n - r & ... \\ p_1 & p_2 & ... & p_n & ... \end{pmatrix}$$

defines another random variable, denoted by $X - r$. Its expectation is obviously $E(X - r) = E(X) - r$.

In general, if X is a random variable with discrete domain

$$D = \{x_1, x_2, ..., x_n, ...\}$$

and $\{p_n\}$ is a distribution associated to X by

$$p_n = P(X = x_n),$$

then the expectation $E(\phi)$ is defined for each function

$$\phi : D \rightarrow \mathbf{R}$$

for which the series $\sum_n \phi(x_n) \cdot p_n$ is a real number. Namely,

$$E(\phi) = \sum_n \phi(x_n) \cdot p_n .$$

In particular, if $\phi(x_n) = \alpha \cdot x_n + \beta$ with "constant" $\alpha, \beta \in \mathbf{R}$, and $E(X)$ exists, then $E(\phi)$ is denoted, by convention, $E(\alpha X + \beta)$ and we have

$$E(\alpha X + \beta) = \alpha \cdot E(X) + \beta .$$

As another particular case, if $\phi(x_n) = x_n^2$ for each $x_n \in D$, then $E(\phi)$ is denoted, by convention, $E(X^2)$.

Proposition 3.4. Under the notations above, if $E(X^2)$ exists, then $E(X^2) \geq E(X)^2$.

Indeed, $E(X^2) - E(X)^2 = \sum_n x_n^2 \cdot p_n - \left(\sum_n x_n \cdot p_n \right)^2$

$$= \sum_n \left(x_n - \sum_m x_m \cdot p_m \right)^2 \cdot p_n \geq 0.$$

Hence the expectation $E(X^2)$ of the square X^2, provided it exists, is always bigger than the square $E(X)^2$. The difference between these numbers is called the **variance** of the random variable X and is denoted $Var(X)$. Therefore,

$$Var(X) = E(X^2) - E(X)^2.$$

Another formula exists

$$Var(X) = E((X - E(X))^2),$$

which expresses the fact that the variance of X is the expectation of the square of the deviation of the random variable X from its expectation $E(X)$. This justifies the use of the variance as a measure of the spread of values around the "center" $E(X)$.

Consider now the elements of the event space Ω are pairs $(x_n, y_m) \in \mathbf{R}^2$, where $\{x_1, x_2, ..., x_n, ...\}$ and $\{y_1, y_2, ..., y_m, ...\}$ are both discrete. If for each pair (x_n, y_m) a number $r_{nm} \geq 0$ is specified, such that

$$\sum_n \sum_m r_{nm} = 1$$

we have a **bi-variate discrete distribution**.

Consider a variable X with discrete domain $D_X = \{x_1, x_2, ..., x_n, ...\}$ and another variable Y with discrete domain $D_Y = \{y_1, y_2, ..., y_m, ...\}$. Given a bi-variate discrete distribution $\{r_{nm}\}$ associated to the pair (X, Y) by

$$r_{nm} = P(X = x_n \cap Y = y_m),$$

the expectation is defined for each real function

$$\Phi : D_X \times D_Y \to \mathbf{R}$$

for which the series $\sum_n \sum_m \Phi(x_n, y_m) \cdot r_{nm}$ is a real number. Obviously,

$$E(\Phi) = \sum_n \sum_m \Phi(x_n, y_m) \cdot r_{nm} .$$

The above considerations apply in particular for the functions "sum" $\Sigma(x_n, y_m) = x_n + y_m$ and "product" $\Pi(x_n, y_m) = x_n \cdot y_m$. Hence

$$E(\Sigma) = \sum_n \sum_m (x_n + y_m) \cdot r_{nm} \text{ and } E(\Pi) = \sum_n \sum_m (x_n \cdot y_m) \cdot r_{nm} .$$

The events $\{Y = y_m\}$ form a countable family; they are mutually exclusive and exhaustive. Hence

$$P(X = x_n) = \sum_m P(X = x_n \cap Y = y_m).$$

Therefore, the variable X is turned on a discrete random variable, by

$$p_n = \sum_m r_{nm} ,$$

and, as such, may have the expectation $E(X)$. Analogously, the variable Y is turned on a discrete random variable by

$$q_m = P(Y = y_m) = \sum_n r_{nm}$$

and may have an expectation $E(Y)$.

Proposition 3.5. Under the notations above, if $E(X)$ and $E(Y)$ exist, then $E(\Sigma)$ exists and $E(\Sigma) = E(X) + E(Y)$.

Let us express Y in table form

$$Y : \begin{pmatrix} y_1 & y_2 & \cdots & y_m & \cdots \\ q_1 & q_2 & \cdots & q_m & \cdots \end{pmatrix}$$

The function Σ is identified to the sum $X + Y$, which is a uni-variate random variable, whose values are the possible distinct sums $x_n + y_m$. Therefore, the property of the expectation E with respect to the sum of random variables, is resumed by the following formula:

$$E(X + Y) = E(X) + E(Y).$$

If the product Π is identified to the product $X \cdot Y$, which is a uni-variate random variable whose values are the possible distinct sums $x_n \cdot y_m$, then the expectations $E(X \cdot Y)$, $E(X)$ and $E(Y)$ may appear. The **covariance** of the pair (X, Y) is defined as

$$Cov(X, Y) = E(X \cdot Y) - E(X) \cdot E(Y).$$

Two discrete random variables X with domain $D_X = \{x_1, x_2, ..., x_n, ...\}$ and Y with domain $D_Y = \{y_1, y_2, ..., y_m, ...\}$ – both uni-variate – are called **independent** if

$$P(X = x_n \cap Y = y_m) = P(X = x_n) \cdot P(Y = y_m)$$

$$\text{for every } x_n \in D_X, \ y_m \in D_Y.$$

An immediate computing proves the following.

Proposition 3.6. If X and Y are independent uni-variate discrete random variables, then $Cov(X, Y) = 0$.

Suppose $Var(X) > 0$ and $Var(Y) > 0$. The relation

$$\alpha^2 Var(X) + 2\alpha Cov(X, Y) + Var(X) = Var(\alpha X + Y) \geq 0$$

is valid for every $\alpha \in \mathbf{R}$. The quadratic polynomial in α cannot have real roots; hence its discriminant is negative. Therefore

$$Cov(X, Y)^2 - Var(X) \cdot Var(Y) \leq 0$$

which is equivalent to

$$-1 \leq \frac{Cov(X, Y)}{\sqrt{Var(X)} \cdot \sqrt{Var(Y)}} \leq 1.$$

The expression

$$\rho(X, Y) = \frac{Cov(X, Y)^2}{\sqrt{Var(X)} \cdot \sqrt{Var(Y)}}$$

is known as the correlation coefficient between X and Y.

In general,

$$Var(X + Y) = Var(X) + Var(Y) + Cov(X, Y).$$

However, if the random variables X and Y are independent, then we have the equality

$$Var(X + Y) = Var(X) + Var(Y).$$

3.11   Examples of Discrete Distributions

A random variable that takes only one value is assimilated to a real number. Genuine random variables take at least two values.

The simplest random variables are the so-called **Bernoulli distributions**, which are associated to some experiments in which the result can be only a "success" or a "failure".

A Bernoulli distribution takes only two values, denoted by:

> 0, which corresponds to "failure", respectively
> 1, which corresponds to "success".

This set of random variable plays an important role in Bayesian networks (see Chapter 6), because the values 0, respectively 1 are easily identified with "false", respectively "true".

A Bernoulli random variable is perfectly determined by the probability of "success"; if this probability is denoted by π, then the corresponding table is:

$$\begin{pmatrix} 0 & 1 \\ 1-\pi & \pi \end{pmatrix}.$$

This particular random variable will be denoted by $Be(\pi)$. Its expectation and variance are easily computed:

$$E(Be(\pi)) = \pi, \ Var(Be(\pi)) = \pi(1-\pi).$$

When throwing a (perfectly equilibrated) coin, the result is expressed by the random variable $Be(\frac{1}{2})$; the values of the random variable $Be(\frac{1}{2})$ are interpreted for example as the number of heads that appear (0 or 1).

The following table

$$\begin{pmatrix} \text{Female} & \text{Male} \\ 0.51 & 0.49 \end{pmatrix}$$

describes the sex of a (future) newborn child. (From statistical data it is known that in every 100 newborn children, usually 51 are girls). Thus the number of boys that are born (in single birth) is expressed as $Be(0.49)$.

Consider four coins that are thrown simultaneously. The number of heads that appear is represented by a random variable, which obviously has only 0, 1, 2, 3 or 4 as values. An immediate computing identifies the respective probabilities:

$$\begin{pmatrix} 0 & 1 & 2 & 3 & 4 \\ \frac{1}{16} & \frac{4}{16} & \frac{6}{16} & \frac{4}{16} & \frac{1}{16} \end{pmatrix}.$$

Let us imagine now that we are not throwing simultaneously four coins, instead we throw four times in sequence the same coin. It is obvious that the "number of heads that appear" is the same random variable as above.

This example admits the following generalization. Consider a sequence $X_1, X_2, ..., X_n$ of n Bernoulli random variables, all of type $Be(\pi)$, each one representing an independent trial. All are characterized by the same probability π of a "success" in the respective trial. The number of overall "successes" in the n consecutive trials is the sum

$$X_1 + X_2 + ... + X_n,$$

which constitutes a new random variable. This is denoted by $b(n, \pi)$, and its values are the numbers: 0, 1, ... , n. The probability to obtain a number of k "successes" in the n consecutive trials – i.e., that variable $b(n, \pi)$ takes k as value – has the following expression

$$P(b(n, \pi) = k) = \binom{n}{k} \pi^k (1 - \pi)^{n-k},$$

thus the distribution table of $b(n, \pi)$ is:

$$\begin{pmatrix} 0 & 1 & ... & k & ... & n \\ (1 - \pi)^n & n\pi(1 - \pi)^{n-1} & ... & \binom{n}{k} \pi^k (1 - \pi)^{n-k} & ... & \pi^n \end{pmatrix}.$$

The random variables $b(n, \pi)$ that are obtained for different numbers n of components and for different probabilities π are called **binomial distributions**.

The values of the binomial distribution $b(n, \pi)$ are discrete; hence the following formula, that gives the "cumulated" probabilities, is obvious:

$$P(b(n, \pi) \le j) = \sum_{k=0}^{j} P(b(n, \pi) = k) \text{ for } j \ge 1.$$

If these values are known, any probability $P(b(n, \pi) = k)$ is immediately obtained:

$$P(b(n, \pi) = k) = P(b(n, \pi) \le k) - P(b(n, \pi) \le k - 1).$$

Another useful formula is the following

$$P(i \le b(n, \pi) \le j) = P(b(n, \pi) \le j) - P(b(n, \pi) \le i - 1).$$

Let us resume: a binomial distribution is linked to an experiment satisfying the conditions:

a) It consists of a number n of trials.
b) The result of each trial is classified either as a "success", or as a "failure".
c) The probability p of a success is the same in all trials.
d) Every trial is independent from each other.

Its values are the number of successes obtained in the n trials.

The expectation of the binomial distribution $b(n, \pi)$ is easily obtained:

$$E(b(n, \pi)) = np .$$

As for the variance, its formula is immediate:

$$Var(b(n, \pi)) = n\pi(1 - \pi) .$$

Example. A test is assembled of 15 questions, each one with five possible answers (and only one correct). Evaluate the probability that a person, who answers at random, will obtain exactly 8 correct answers (i.e. over 50% success rate). Compute the average of correct answers obtained by persons who answer at random.

When answering at random, the number of correct answers is a binomial random variable (one says also that "is binomially distributed") associated to a number of 15 trials, and the probability of success, in each trial, is $\pi = \dfrac{1}{5} = 0.2$. If the number of correct answers is $k = 8$, then we should look in the tables for (or to compute) the probability $P(b(15, \ 0.2) = 8)$, which is 0.0035, hence less than 1%!

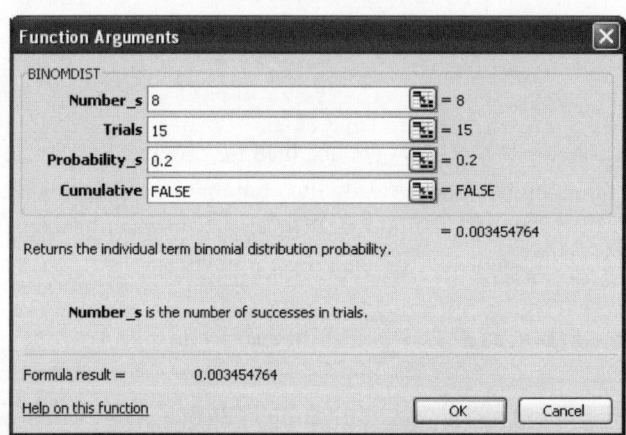

Fig. 3.9. Binomial distribution in *Excel*

On the other hand,

$E(b(15, \ 0.2)) = 15 \cdot 0.2 = 3$.

Therefore, a number of 3 correct answers will be obtained in average.

There exist tables that contain the values $P(b(n, \pi) = k)$ for "standardized" values of π ($\pi = 0.25$, $\pi = 0.1$ and others). However, the present use of computers makes these tables superfluous.

In *Microsoft Excel* there is a built-in function BINOMDIST with four arguments (see Figure 3.9 above):

— Number of "successes" k.
— Number of trials n.
— Probability of a success π.
— A logical parameter, whose value TRUE indicates the cumulated probability.

In *Microsoft Excel* there is another built-in function NEGBINOMDIST. This is used to obtain the probability of exactly f failures before obtaining the kth success (on condition that the probability p of a success is the same for all trials).

Another family of discrete random variables is that of Poisson[6] distributions. As pointed out above, a binomial random variable counts the number of "successes" obtained in a fixed number, n, of trials. Similarly, a Poisson random variable counts the number of "rare appearances" that happen in a given time interval, or in a delimited space interval (a region, a domain).

A Poisson experiment is characterized by three conditions:

1) The number of appearances in a given interval is independent to what happens into any other interval.
2) The probability of a single appearance in a given interval is proportional to the "length" of that interval.
3) The probability of several appearances in an interval tends to 0 when the "length" of the interval tends to 0 (this is interpreted, as "the appearances are rare").

A **Poisson distribution** represents the "chances" of a number of appearances that happen in a given interval, when the conditions of a Poisson experiment are met. Such a distribution depends on a single (real positive) parameter λ and is usually denoted as $Po(\lambda)$. It takes as values the natural numbers 0, 1, 2, ..., n, ... with respective probabilities

$$P(Po(\lambda) = n) = \frac{\lambda^n}{n!} \cdot \exp(-n).$$

Several tables that contain the values of probabilities $P(Po(\lambda) = n)$ for different particular values of the parameter λ exist. However, such tables are superfluous today; for example, in *Microsoft Excel* the built-in function POISSON is available.

It is relatively easy to establish the following results concerning the expectation and the variance of a Poisson distribution:

$$E(Po(\lambda)) = \lambda \text{ and } Var(Po(\lambda)) = \lambda.$$

Examples. 1) In biology and other life sciences it is accepted that the incidence of attacks of parasites on a population is well described by a Poisson distribution. Here n is the number of parasites that attack the same individual of the population.

2) The chief of an emergency unit knows, from his past experience, that in average 12 emergency calls are received each month that impose the use of the helicopter.

---

[6] Denis Poisson (1781-1840) – French mathematician and physicist, creator of the mathematical physics.

However, the helicopter is able to fulfill at most three missions per day. What is the probability that, in an ordinary day, more that 3 emergency calls that impose the use of the helicopter are received?

The estimation is simple, since such phenomena are modeled using Poisson distributions. In our case $\lambda = \dfrac{12}{30} = 0.4$ (let us admit every month has 30 days). The probability of an overflow

$$P(Po(\lambda) > 3) = 1 - P(Po(\lambda) \le 3) = 0.00077$$

is extremely low. However, once in three years we should expect an exceptional situation!

3.12   Continuous Distributions

In previous sections only random variables whose domain (i.e. the set of values) was a discrete subset of $\mathbf{R}$ were treated. We will consider now (uni-variate) random variables whose domain is assimilated to the whole set $\mathbf{R}$ of real numbers. These are known as the **continuous distributions**.

Values of such a distribution appear, obviously, as results of measurements (lengths, weights, time durations, temperatures, concentrations etc.).

For continuous distributions it is impossible to say, in general, that they describe the distribution of "chances" among the different possible values of the corresponding random variables. In fact, if X is a continuous random variable, then the probability of the "event" $X = x$ is, in general, zero!

A continuous distribution is described by a real function

$$F : \mathbf{R} \to [0,1],$$

which satisfies the following conditions:

(Di1) $F(x_1) \le F(x_2)$ for every pair of real numbers $x_1 < x_2$.

(Di2) $\lim\limits_{x \to -\infty} F(x) = 0$ and $\lim\limits_{x \to \infty} F(x) = 1$

and the connection with the random variable X is given by the relation

$$P(X \le x) = F(x) \text{ for every } x \in \mathbf{R}.$$

The monotone increasing function F is called the **distribution function** of the random variable X.

If we follow this approach, then any discrete distribution is in fact nothing else than a particular case of continuous distributions. Indeed, if

$$x_1 < x_2 < ... < x_n < ...$$

is the ordered sequence of values of the discrete random variable X (that is associated to the discrete distribution), then the diagram of cumulated (relative) frequencies is in fact a graphical representation of the distribution function. For a number x inside the interval (x_k, x_{k+1}), it is obvious that

$$F(x) = P(X \leq x) = P(X \leq x_k) = F(x_k),$$

which explains the "stepwise" aspect of the distribution function (see Figure 3.10 below).

The most important distribution functions are described by use of the so-called density functions.

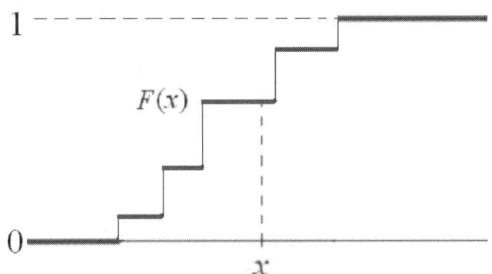

Fig. 3.10. Distribution function of a discrete random variable

A **density function** is a real function

$$f : \mathbf{R} \to \mathbf{R},$$

which satisfies the following two conditions:

(De1) $f(x) \geq 0$ for every $x \in \mathbf{R}$,

(De2) $\int_{-\infty}^{\infty} f(x)\, dx = 1$.

The connection between a density function and a distribution function, hence a random variable X, is given by the following relations:

$$F(x) = P(X \leq x) = \int_{-\infty}^{x} f(x)\, dx \text{ and } f(x) = \frac{dF}{dx}(x),$$

provided f is a continuous function. Moreover,

$$P(a < X \leq b) = \int_{a}^{b} f(x)\, dx.$$

As in the discrete case, for continuous distributions the expectation and the variance are defined. More precisely, once a density function φ is specified, the expectation $E(\phi)$ is defined for each real function

$$\phi : \mathbf{R} \to \mathbf{R}$$

for which the integral $\int_{-\infty}^{\infty} \phi(x) f(x)\, dx$ is a real number. As an important particular case, for the square $\phi(x) = x^2$, if $E(\phi)$ exists, it will be denoted $E(X^2)$. As in the discrete case, $E(X^2) \geq E(X)^2$, and the variance of X is defined as $Var(X) = E(X)^2 - E(X)^2 = E((X - E(X))^2)$.

Of course, the expectation $E(X)$, if it exists, is a "center" of the values of X. The variance $Var(X)$ is a measure of the spread of values around the center.

The main properties of the expectation and variance cannot be obtained without recurring to multi-variate random variables.

The simplest case is that of bi-variate random variables. Their domain is the plane $\mathbf{R}^2$ of pairs of real numbers.

A continuous bi-variate distribution is described by a real function

$$F : \mathbf{R}^2 \to [0,1]$$

satisfying the following six conditions:

(2Di1) $F(x_1, y) \leq F(x_2, y)$ for every $x_1 < x_2$ and y.

(2Di2) $F(x, y_1) \leq F(x, y_2)$ for every x and $y_1 < y_2$.

(2Di3) $F(x_1, y_1) - F(x_1, y_2) - F(x_2, y_1) + F(x_2, y_2) \geq 0$

for every $x_1 < x_2$ and $y_1 < y_2$.

(2Di4) $\lim_{x \to -\infty} F(x, y) = 0$ for every y.

(2Di5) $\lim_{y \to -\infty} F(x, y) = 0$ for every x.

(2Di6) $\lim_{x,y \to +\infty} F(x, y) = 1$.

Of course, a pair (X, Y) of (uni-variate) random variables X and Y could represent a bi-variate random variable.

As such, the connection between the distribution F and the random variable (X, Y) is expressed in the following way

$$F(x, y) = P(X \leq x \cap Y \leq y) \text{ for every } (x, y) \in \mathbf{R}^2.$$

The most interesting situation appears when

$$F(x, y) = \int_{-\infty}^{y} \int_{-\infty}^{x} f(u, v)\, du\, dv$$

where $f : \mathbf{R}^2 \to \mathbf{R}$ is a two-variable function with positive values, called the **joint density function**.

In case the joint density function f is specified, the expectation $E(\Phi)$ is defined for each real function

$$\Phi : \mathbf{R}^2 \to \mathbf{R}$$

for which the integral $\int_{-\infty}^{\infty} \int_{-\infty}^{\infty} \Phi(x, y) f(x, y) \, dx \, dy$ is a real number.

The main particular cases are:

- The affine functions $\Lambda(x, y) = \alpha \cdot x + \beta \cdot y + \gamma$ with "constant" $\alpha, \beta, \gamma \in \mathbf{R}$, and in particular

- The sum $\Sigma(x, y) = x + y$ and the mean $A(x, y) = \dfrac{1}{2} \cdot x + \dfrac{1}{2} \cdot y$,

- The product $\Pi(x, y) = x \cdot y$.

As in the discrete case, the additivity property of the expectation:

$$E(X + Y) = E(X) + E(Y)$$

can be established. Moreover, the **covariance** of the pair (X, Y) is defined as

$$Cov(X, Y) = E(X \cdot Y) - E(X) \cdot E(Y).$$

Suppose the uni-variate random variables X, Y are described, respectively, by the distribution functions

$$F, G : \mathbf{R} \to [0, 1].$$

This means we have the relations

$$P(X \leq x) = F(x), \ P(Y \leq y) = G(y) \text{ for every } x, y \in \mathbf{R}.$$

The possible values x, y of the random variables could be added. Based on the addition of values, we may define the sum $X + Y$ of random variables in case we are able to compute

$$P(X + Y \leq s) = H(s) \text{ for every } s \in \mathbf{R}.$$

Of course, the sum $X + Y$ will be described by the distribution function $H : \mathbf{R} \to [0, 1]$.

The following Proposition is a standard result in classical Probability Theory.

Proposition 3.8. If the joint density of the pair (X, Y) is

$$f : \mathbf{R}^2 \to \mathbf{R}$$

and if $S = X + Y$, then the density function of S exists and is given by:

$$h(s) = \int\limits_{-\infty}^{+\infty} f(u, s - u) \, du \ .$$

Indeed, for any real number s we have

$$P(S \le s) = P(X + Y \le s) = \iint\limits_{u+v \le s} f(u, v) \, du \, dv \ .$$

A simple change of variable v, in the integral, to $w = u + v$ leads to

$$P(S \le s) = \int\limits_{-\infty}^{s} \int\limits_{-\infty}^{+\infty} f(u, w - u) \, du \, dw$$

which, compared to $\int\limits_{-\infty}^{s} h(w) \, dw$, proves the Proposition.

Analogous results can be established for the product $P = X \cdot Y$, the quotient $Q = X / Y$ and other random variables (that are defined accordingly).

3.13 Examples of Continuous Distributions. Normal

In many cases of theoretical reasoning normal (i.e. Gaussian) distributions play important roles. A **normal distribution**, determined by the real parameters μ and $\sigma^2 > 0$, is denoted by $N(\mu, \sigma^2)$ and is characterized by the density function:

$$\varphi(x) = \frac{1}{\sigma\sqrt{2\pi}} \exp\left(-\frac{(x - \mu)^2}{2\sigma^2} \right).$$

The graph of this function is bell-shaped – usually known as "the Gauss' bell" – and is symmetric with respect to the vertical line $x = \mu$.

After computing some integrals the following results are obtained

$$E(N(\mu, \sigma^2)) = \mu \text{ and } Var(N(\mu, \sigma^2)) = \sigma^2,$$

formulas that give an obvious interpretation of the two parameters. The parameter μ is referred to as the (theoretical) mean, and σ^2 is referred to as the (theoretical) variance, i.e. the square of the (theoretical) standard deviation $\sigma > 0$ of the random variable $N(\mu, \sigma^2)$. It is not surprising that the "spread of the bell" depends on the extent of σ (see Figure 3.11 below for some examples).

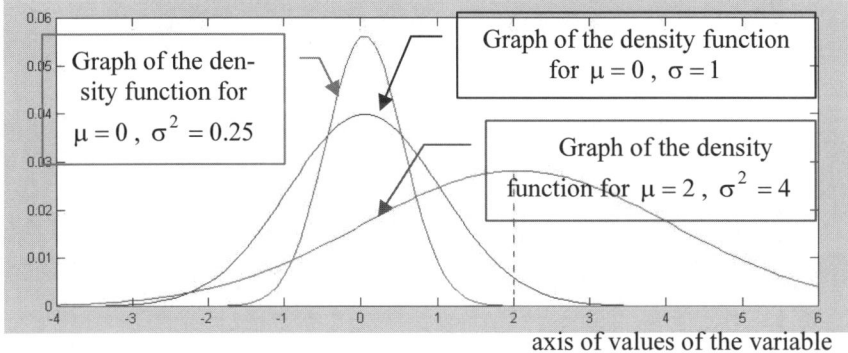

axis of values of the variable

Fig. 3.11. Examples of normal densities

Obviously, a normal distribution may take any real number as a value. However, the "probability of values" diminishes as they depart "away" from the mean μ.

It is impossible to fill in tables for all normal distributions. The following result was extensively used in the past to compute in relation to normal distributions. If X is a distribution of type $N(\mu, \sigma^2)$, then the distribution

$$Z = \frac{1}{\sigma}(X - \mu)$$

is of type $N(0, 1)$. Moreover,

$$P(X \le x) = P\left(Z \le \frac{x - \mu}{\sigma}\right)$$

and this allows us to use data found in tables of the so-called standard normal distribution $N(0, 1)$.

However, the use of tables is obsolete today, and the general software allows all kind of computing related to normal distributions. For example, in *Excel* two built-in functions, called NORMDIST and NORMINV, which depend on the parameters μ and σ, are available. The dialog box of the first is presented in the Figure 3.12 below. To compute values in the classical way, i.e. in relation to the standard normal distribution, two supplementary built-in functions NORMSDIST and NORMSINV are also available.

A (bell shaped) Gaussian density function that corresponds to a continuous distribution, is ideal, it cannot appear in connection with a natural population. However, some natural "biological" distributions – such as the distribution of the height (in cm) of humans, or the distribution of the intelligence quotient (IQ) of humans – are approximately bell-shaped. (For example, the IQ of humans is assimilated to a normal distribution of mean 100 and standard deviation 10.)

Traditionally, normal distributions quantify the involuntary errors that appear in measurements of lengths or weights. Here, if μ represents the measured value, σ

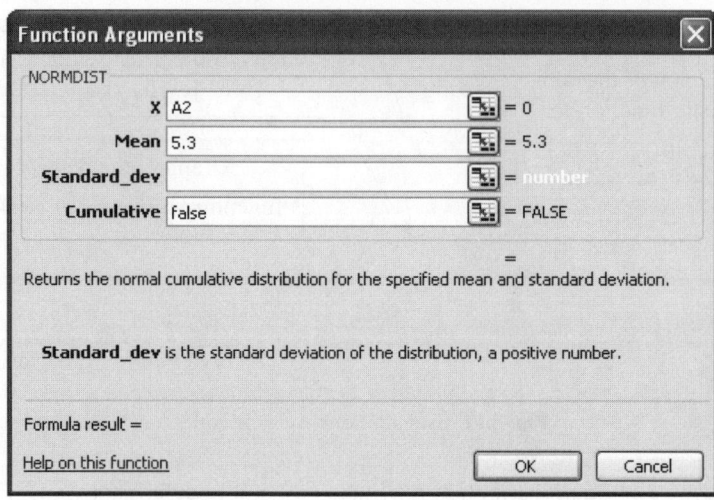

Fig. 3.12. Computing with normal distributions in *Excel*

will represent the measuring error. Normal distributions are also assimilated to the so-called "noise" that affects data transmission.

Sometimes the relative position of an individual from a normally distributed population is of interest. For example, we know that an individual has obtained a result of 80 in a competition. This number, 80, tells us nothing about the classification of that individual, because it can be near to the minimum result, near to the median, as well as near to the maximum. The real performance is not obvious!

A usual method to describe the real performance is to indicate the standard score (known also as the z-score). This score expresses how much standard deviation is "under" the result. It is computed easily, by subtracting the population mean μ from the result x, then by diving the difference $x - \mu$ to the standard deviation σ.

$$z = \frac{x - \mu}{\sigma}.$$

In practice μ and σ are estimated from the data available. For example, if the result 80 was obtained by a person at an IQ test (for which it is assumed that $\mu = 100$ and $\sigma = 10$, then the z-score has value –2, which corresponds to a weak performance of that person.

The normal distribution helps us to define what a "normal" individual means. In quality theory, individuals having z-scores between –2 and +2 are labeled as "standard", and individuals whose z-scores are between –3 and +3 are labeled as "normal".

Let us notice that, by this **standardization**, a value of an arbitrary normal distribution is replaced by a value of the standard normal distribution $N(0,1)$. Keep in mind the idea that by standardization the scores of (individuals from) different populations could be compared.

The family of normal distribution is important because some very interesting results, concerning arbitrary distributions, are expressed in terms of normal distributions. For the theory of sampling the following result is extremely important: if $X_1, X_2, ..., X_n$ are distributions of type $N(\mu, \sigma^2)$, i.e. are normal with the same theoretical mean μ and the same theoretical standard deviation σ, then the distribution of the average

$$Y = \frac{1}{n}(X_1 + X_2 + ... + X_n)$$

is also normal, with the same mean μ and another standard deviation (smaller than σ). It is established that this standard deviation is exactly $\sigma / \sqrt{n}$ in case all $X_1, X_2, ..., X_n$ are mutually independent. In measure theory the above result is interpreted as follows: the distribution X expresses the result of a single measurement, the parameter μ expresses the obtained value, and the parameter σ expresses the inverse of the precision. (High precision means σ small, near to 0.) After a series of measurements $X_1, X_2, ..., X_n$ (of the same object), we expect the average to express the measure of that object, with higher precision.

Normal distributions appear in several theoretical results that express other kind of "limit behavior" of random variables of other types. For example, if we consider a binomial distribution $b(n, p)$ where the parameter n is "big" and the parameter p is not near to the extremes 0 and 1, then the histogram of its probabilities will suggest a Gaussian bell. This allows the use of the normal distribution as an approximation of such a binomial $b(n, p)$. More precisely, the normal $N(\mu, \sigma^2)$ having the mean $\mu = np$ and the variance $\sigma^2 = np(1 - p)$ is a good approximation. Thus, calculus of the value $P\left(k - \frac{1}{2} < N(\mu, \sigma^2) \le k + \frac{1}{2} \right)$ replaces the direct calculus of the value $P(b(n, p) = k)$, which is more tedious.

Let us mention that the approximation of a binomial distribution with a normal distribution is satisfactory when:

$$0.1 < p < 0.9, \quad n > 30 \quad \text{and} \quad np > 5.$$

3.14 Examples of Continuous Distributions. Chi-Square

It was stated above that, by using the "standardization formula"

$$Z = \frac{X - \mu}{\sigma}$$

the normal distribution $X \in N(\mu, \sigma^2)$ is replaced by the "normal standard" distribution Z (i.e. $Z \in N(0, 1)$). The values of Z are, exactly as those of X, either positive or negative; however, the former are "grouped" symmetrically around the origin. (In fact, 99.7% of the values are situated between −3 and +3.)

What can be said about the square Z^2 ? Of course, its values cannot be negative, thus they are definitely not symmetrical around the origin. (Still, 99.7% of the values are between 0 and $9 = 3^2$!)

The density function of the square Z^2 is easily represented; by using the function CHIDIST implemented in *Excel*. In the Figure 3.13 below it is apparent that no "axis of symmetry" exists, and the expectation is not obvious.

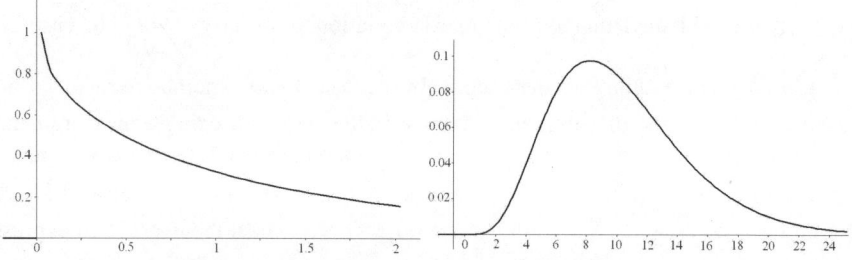

Fig. 3.13. Density function of Z^2 (left) and of $\chi^2(10)$

When is the square Z^2 useful? Of course, if Z expresses a random error (positive or negative), then Z^2 will express the square of the error. In some situations, after a series of measurements (of the same object), the "cumulative" error is expressed as a sum of squares, not necessarily reduced at a singe term. This imposes the following generalization. Consider several random variables $Z_1, Z_2, ..., Z_\nu$ (all normal standard and mutually independent). The sum of squares

$$Z_1^2 + Z_2^2 + ... + Z_\nu^2$$

considered as a random variable, is known as the **chi-square distribution** with ν degrees of freedom and is denoted $\chi^2(\nu)$.

Therefore, we have a new family $\{\chi^2(\nu)\}$ of continuous random variables, which "depend" on the parameter ν, the number of degrees of freedom, which is a natural number (1, 2, 3 etc.).

The function CHIDIST from *Excel* allows us to represent the density functions of these random variables – see the Figure 3.13 above for examples and Figure 3.14 below for the control.

These densities are uni-modal, the unique peak being above the abscissa $\nu - 2$. It is easy to understand the topics

CHIDIST(abscissa x, number of degrees of freedom).

However, the use of CHIDIST in *Excel* is different from that of NORMSDIST. For historical reasons, in fact CHIDIST(x, v) is exactly the probability that the distribution $\chi^2(v)$ takes values larger than x (and not less than x as in the case of NORMSDIST).

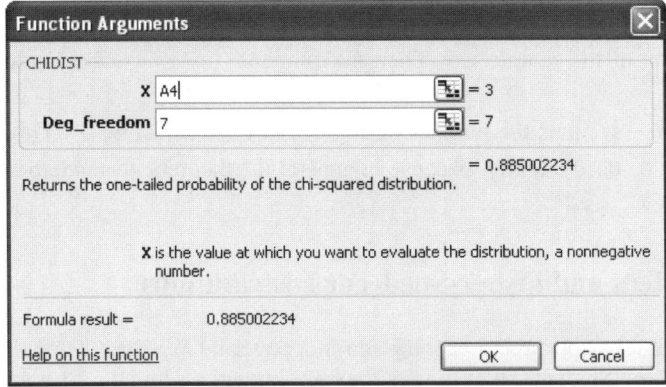

Fig. 3.14. Use of CHIDIST in *Excel*

Computation allows us to obtain the expectation and the variance of these distributions. The results are easy to remember, being connected to the number of degrees of freedom:

$$E(\chi^2(v)) = v \, , \; Var(\chi^2(v)) = 2v \, .$$

The intensive use of the family of chi-square distributions $\{\chi^2(v)\}$ is due to important results in sampling theory. One of these results is expressed as follows. Suppose our individuals under study are grouped in several "modalities" (taking into account two criteria); the respective numbers are inserted in a contingency table

		Column c		Row totals
	...	...	...	...
Row r	...	n_{rc}	...	$n_{i\bullet}$
	...	...	...	...
Column totals	...	$n_{\bullet c}$	...	$n_{\bullet\bullet}$

Consider the statistic

$$X^2 = \sum_r \sum_c \frac{n_{\bullet\bullet}}{n_{r\bullet} \times n_{\bullet c}} \left(n_{rc} - \frac{n_{r\bullet} \times n_{\bullet c}}{n_{\bullet\bullet}} \right)^2 .$$

This statistic is following approximately the distribution $\chi^2(\nu)$, where the number of degrees of freedom is exactly $(R-1)\cdot(C-1)$, R being the number of rows, C the number of columns of the data table.

In sampling theory we encounter another interesting result. Suppose we have a sample of volume n from a normally distributed population $N(\mu,\sigma^2)$ and the standard deviation of the sample is s. Then the quotient $\dfrac{(n-1)s^2}{\sigma^2}$ is a random variable of type $\chi^2(n-1)$, thus the number of degrees of freedom is $n-1$. (Obviously, this result cannot be used directly, because in general the theoretical standard deviation σ is not known!)

3.15   Student and Fisher-Snedecor Distributions

The t-distributions appeared a century ago in a paper by W. Gosset[7], published under the pseudonym "Student", that is why they are known also as the **Student distributions**. The family $\{t(\nu)\}$ of these distributions is "parameterized" by the same number ν as the family $\{\chi^2(\nu)\}$. The reason is obvious if we take into account the definition:

$$t(\nu) = \frac{Z}{\sqrt{\chi^2(\nu)/\nu}}\,.$$

Remember that Z represents a standard normal distribution, i.e. of type $N(0,1)$, and $\chi^2(\nu)\big/\nu$ is the arithmetic mean of squares of ν copies of distributions of type $N(0,1)$.

To compute with this family $\{t(\nu)\}$ of distribution, in *Excel* a built-in function TDIST is available. Its use is as follows.

TDIST(abscissa x, number of degrees of freedom, laterality parameter).

The laterality parameter has values either 1 (i.e. unilateral) or 2 (i.e. bilateral).

In the Figure 3.15 below the density function of the distribution t(10) is presented. It is apparent that the "curve" is symmetric with respect to the origin, its graph is "similar" – however, not identical – to that of a Gaussian. (In fact, for $\nu > 30$ the graphs of $t(\nu)$ and $N(0,1)$ practically coincide.)

The use of the family $\{t(\nu)\}$ of distributions is due also to some important results from the sampling theory. Two of these results are as follows.

Suppose the population is normally distributed, with mean μ and variance σ^2, and suppose $x_1, x_2,..., x_n$ is a "small" sample of volume n. Denote by

---

[7] William Sealey Gossett (1876-1937), English mathematician.

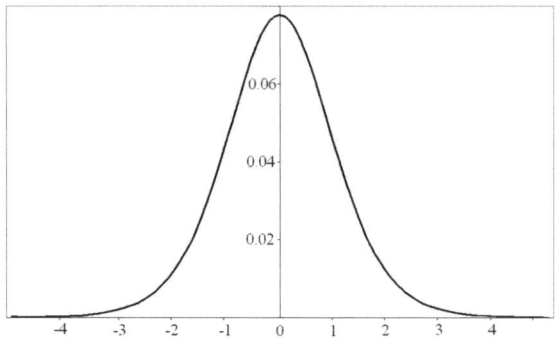

Fig. 3.15. Density function of t(10)

$$m = \frac{x_1 + x_2 + \dots + x_n}{n}$$ the sampling mean and by s the sampling standard deviation

$$\sqrt{\frac{(x_1 - m)^2 + \dots + (x_n - m)^2}{n-1}}$$. Then the quotient $\dfrac{m - \mu}{s \big/ \sqrt{n}}$ follows a distribution of

type $t(n-1)$ with $n-1$ degrees of freedom.

(Let us mention here that the quotient $\dfrac{(n-1)s^2}{\sigma^2}$ follows a chi-square distribution

with $n-1$ degrees of freedom.)

Suppose two samples from the same (normally distributed) population are available. From the first sample, of volume n_1, the sample mean m_1 and the sample standard deviation s_1 are computed. Analogously, from the second sample, of volume n_2, the sample mean m_2 and the sample standard deviation s_2 are computed.

If at least one of the numbers n_1, n_2 is "small", then the quotient $\dfrac{m_1 - m_2}{s\sqrt{\dfrac{1}{n_1} + \dfrac{1}{n_2}}}$ is

Student distributed with $n_1 - n_2 - 2$ degrees of freedom. In the quotient above the square s^2 of s is a weighted average of the squares s_1^2 and s_2^2 of the respective standard deviations, more precisely

$$s^2 = \frac{(n_1 - 1)s_1^2 + (n_2 - 1)s_2^2}{n_1 + n_2 - 2}.$$

The use of the family of distributions $\{F(v_1, v_2)\}$, which are known as **Fisher-Snedecor**[8] **distributions**, is also due to some results from the sampling theory. The motivation appears when considering "practical" problems of the following type: if data from two samples are available, samples from distinct populations supposed normally distributed, it is true that the spread of individuals in the populations is the same? (In other words, the theoretical standard deviations – or the variances – of the two populations are the same?)

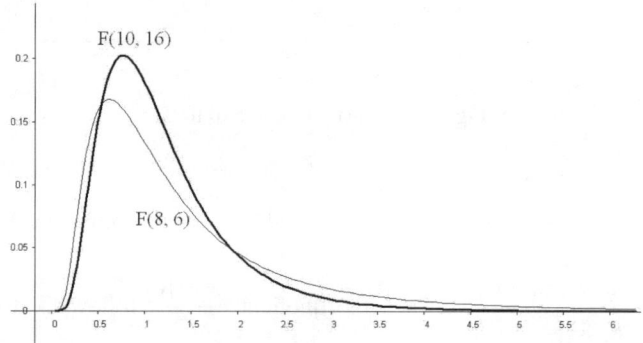

Fig. 3.16. Density functions of F(8, 6) and of F(10, 16)

Obviously, if this is true, then the quotient of the respective variances (or theoretical standard deviations) is 1. Theoretical standard deviations are estimated by sampling standard deviations. When the quotient of the sampling standard deviations is far from 1 (i.e. is either "big" or "near 0"), we will have no reason to accept that variances coincide.

The definition of (theoretical) Fisher-Snedecor distributions is justified by the fact that theoretical standard deviations are estimated by sampling standard deviations. By definition, $F(v_1, v_2)$ is the quotient $\dfrac{\chi^2(v_1)/v_1}{\chi^2(v_2)/v_2}$ of two arithmetic means.

In the Figure 3.16 above the density function of two such distributions is represented. The similarity with density function of chi-square distribution is apparent (?!).

Of course, in *Excel* the built-in function that allows computing with Fisher-Snedecor distributions is FDIST.

In case of two samples of volumes n_1 res. n_2, extracted from normally distributed populations having the same variance – in particular extracted from the same (normally distributed) population, the quotient of the squares of the sampling standard

---

[8] G. W. Snedecor (1881-1974), American statistician.

deviations $\dfrac{s_1^2}{s_2^2}$ follows a distribution of type $F(n_1 - 1, n_2 - 1)$, the number of degrees of freedom being $n_1 - 1$ and $n_2 - 1$.

The "practical" use of all distributions N, χ^2, t, F above (and of many others) in significance testing supposes the choice of a threshold that delimits the "rejection region". This region will contain values that are considered "significantly" different from 0 (in cases of N, χ^2, t) respectively from 1 (in case of F). What "significantly" means is a personal option, related to the risks we are ready to accept. In Chapter 4 this subject will be approached.

3.16 Formal Definition of Random Variables

The treatment of random variables above is applications oriented; a formal definition of (uni- or multi-variate) random variables is given in the context of the so-called Measure Theory. Recall the intervals $[a, b) \subset \mathbf{R}$ generate a sigma-algebra $\mathcal{B}$ over $\mathbf{R}$, which is called the real Borel sigma-algebra, and whose elements are called Borel sets. More generally, the Cartesian products $\overset{m}{\underset{i=1}{\times}}[a_i, b_i)$ generate a sigma-algebra over the m-dimensional real space $\mathbf{R}^m$. The "events" of this sigma-algebra $\mathcal{B}^m$ are called m-dimensional Borel sets.

Let us start with a sigma-algebra $\mathcal{A}$ over the universe Ω.

A function $X : \Omega \to \mathbf{R}^m$ is called **random variable** if

$X^{-1}(B) \in \mathcal{A}$ for any m-dimensional Borel set B.

Here $X^{-1}(B)$ is a notation for the set $\{\omega \in \Omega \mid X(\omega) \in B\}$.

According to this definition, the notion of random variable depends on the sigma-algebra $\mathcal{A}$. In case $\mathcal{A} = 2^{\Omega}$ all functions $X : \Omega \to \mathbf{R}^m$ are random variables.

As an example, suppose we toss a coin several times, each time $h = 1$ denotes the apparition of head, $h = 0$ denotes the apparition of tail. Then $\omega = h_1 h_2 h_3 h_4$ denotes the combination of heads/tails that appear in four consecutive tosses. Ω has 16 elements, and $\mathcal{A} = 2^{\Omega}$ has 65536 elements. $X(\omega) = h_1 + h_2 + h_3 + h_4$, which denotes the number of heads in four consecutive tosses, is a random variable. Also $X'(\omega) = (h_1 + h_2, h_3 + h_4)$ is a (bi-variate) random variable. A less obvious example is X'', where $X''(\omega) = r$, the rank of the first head that appears (or 0 if no head will appear).

Consider another universe Ω' and a sigma-algebra $\mathcal{A}'$ over Ω'. The following definitions are taken from Measure Theory.

The pair $(\Omega, \mathcal{A})$, where $\mathcal{A}$ is a sigma-algebra over Ω, is called **measurable space**. A **measurable function**

$$F : (\Omega, \mathcal{A}) \rightarrow (\Omega', \mathcal{A}')$$

is simply a function $F : \Omega \rightarrow \Omega'$ such that $F^{-1}(A') \in \mathcal{A}$ for any $A' \in \mathcal{A}'$.

Thus, random variables are nothing else than measurable functions $(\Omega, \mathcal{A}) \rightarrow (\mathbf{R}^m, \mathcal{B}^m)$.

In particular, measurable functions

$$Y : (\mathbf{R}^n, \mathcal{B}^n) \rightarrow (\mathbf{R}, \mathcal{B})$$

are simply n-argument functions $Y : \mathbf{R}^n \rightarrow \mathbf{R}$ that satisfy the condition

$Y^{-1}(B) \in \mathcal{B}^n$ for any (real) Borel set B.

Theorem 3.9. Let $\mathcal{A}$ be a sigma-algebra of events over the universe Ω, let $X_1, ..., X_n : \Omega \rightarrow \mathbf{R}$ be (univariate) random variables, and $Y : \mathbf{R}^n \rightarrow \mathbf{R}$ a measurable function. Then the composition

$$Y \circ (X_1, ..., X_n) : \Omega \rightarrow \mathbf{R}$$

given by $Y \circ (X_1, ..., X_n)(\omega) = Y(X_1(\omega), ..., X_n(\omega))$ is a random variable.

The **proof** is standard in Measure Theory (see [König 1997], [Taylor 1997]). In short, denote by B a real Borel set. Then, by hypothesis, $Y^{-1}(B) \in \mathcal{B}^n$.

On the other hand,

$$(Y \circ (X_1, ..., X_n))^{-1}(B) = \{\omega \mid (X_1(\omega), ..., X_n(\omega)) \in Y^{-1}(B)\}.$$

Now, for $D = \underset{i=1}{\overset{n}{\times}} [a_i, b_i)$, we have

$$\{\omega \mid (X_1(\omega), ..., X_n(\omega)) \in D\} = \bigcap_{i=1}^{n} \{\omega \mid X_i(\omega) \in [a_i, b_i)\} \in \mathcal{B}.$$

Sets D generate the sigma-algebra $\mathcal{B}^n$; therefore,

$$\{\omega \mid (X_1(\omega), ..., X_n(\omega)) \in C\} \in \mathcal{B} \text{ for each } C \in \mathcal{B}^n,$$

in particular for $C = Y^{-1}(B)$. Thus,

$$(Y \circ (X_1, ..., X_n))^{-1}(B) \in \mathcal{B}$$

which ends the proof.

As particular cases, $X_1 + X_2$, $X_1 - X_2$, $X_1 \cdot X_2$, $\min\{X_1, X_2\}$, $\max\{X_1, X_2\}$ are random variables, provided X_1, X_2 are random variables.

The set $\mathbf{R}$ of real numbers is totally ordered by $\leq$. As a direct consequence, an ordering between uni-variate random variables $\Omega \to \mathbf{R}$ can be defined by

$$X_1 \leq X_2 \text{ iff } X_1(\omega) \leq X_2(\omega) \text{ for all } \omega \in \Omega.$$

Each real number $a \in \mathbf{R}$ can be assimilated to a "trivial random variable" $a : \Omega \to \mathbf{R}$, defined by $a(\omega) = a$ for all $\omega \in \Omega$. Of course, $X \leq a$ means $X(\omega) \leq a$ for all ω.

Consider a sequence $X_1, X_2, ..., X_n, ...$ of uni-variate random variables. Given $\omega \in \Omega$, a sequence of real numbers

$$X_1(\omega), X_2(\omega), ..., X_n(\omega), ...$$

appears. If for $\omega \in \Omega$ this sequence is convergent, denote by $X(\omega)$ its limit. Thus,

$$X(\omega) = \lim_n X_n(\omega).$$

Suppose for all $\omega \in \Omega$ the sequence above is convergent and all $X(\omega)$ are real numbers. In this case the numbers $X(\omega)$ determine a new random variable X, and it is natural to denote it as $\lim_n X_n$.

We have now all the ingredients needed to formally define the expectation $E(X)$, at least for some uni-variate random variables X.

There are two conditions that the **expectation operator** E is bound to fulfill:

(E1)   For any $a, b \in \mathbf{R}$, if $a \leq X \leq b$, then $a \leq E(X) \leq b$.

(E2)   If $X_1, X_2, ..., X_n, ...$ are random variables such that $X_1 \leq X_2 \leq ...$ $... \leq X_n \leq ...$ and $X = \lim_n X_n$ exists, then $E(X) = \lim_n E(X_n)$.

As an example, consider Ω finite, $\Omega = \{\omega_1, \omega_2, ..., \omega_K\}$. Given positive numbers $p_1, p_2, ..., p_K$ such that $p_1 + p_2 + ... + p_K = 1$, the formula

$$E(X) = \sum_{k=1}^{K} p_k \cdot X(\omega_k)$$

defines an operator E that satisfies conditions (E1) and (E2). An expectation of this kind is called a "weighted average".

Any expectation operator E gives rise to a probability. Indeed, consider an event A from the chosen sigma-algebra $\mathcal{A}$ of subsets of Ω. The characteristic function of A

$$\chi_A(\omega) = \begin{cases} 1 & \text{if } \omega \in A \\ 0 & \text{if } \omega \notin A \end{cases}$$

is a random variable $\Omega \to \mathbf{R}$. It is immediate that $0 \leq \chi_A \leq 1$, hence

$$0 \leq E(\chi_A) \leq 1.$$

Now, the probability of the event A may be defined by

$$P(A) = E(\chi_A).$$

More details are found in [Halmos 1974] or [Doob 1994].

3.17 Probabilities of Formulas

Probability Theory is strictly connected to Measure Theory, where the notions of length, surface and volume are defined and extended. However, what is important is that the results obtained in Probability Theory can be adapted to sentences and reasoning.

As we pointed out in the previous chapter, in the basic Probability Theory the probability is a function defined on events. Such events could be related to a random variable. In these situations statements (formulas) express the events. We will now try to define directly a calculus with probabilities of formulas, taking into account that a statement could be "true" in some worlds and "false" in others.

More precisely, the probability P is a real function with a set ("event") as argument. It is very convenient to take a description of the event, regarding this as equivalent to the set. For example, given random variables X and Y defined over Ω, it is convenient to write $P(X < Y)$ instead of the formally correct $P(\{\omega \in \Omega \mid X(\omega) < Y(\omega)\})$. "$X < Y$" is a formula.

In the following, we figure out an observer receiving messages from the environment. These messages give him hints about the truth status of several statements of interest. The observer imagines several worlds, such that for each pair there exist a statement that is true in one world and false in the other. These are called the possible worlds, and their set will be denoted by Ω. (One of these worlds is the actual one, but the observer does not know which.)

A **measure** of the worlds is a function

$$m : \Omega \to [0,1]$$

such that:

(M1) $m(\omega) \geq 0$ for all $\omega \in \Omega$,

(M2) $\sum_{\omega \in \Omega} m(\omega) = 1$.

(i.e. the measures of all possible worlds are positive and sum to 1)

(For example, the function m could express the belief of our observer about the chances of possible worlds to be the actual one.)

The **probability of a formula** f, written $P(f)$, is the sum of the measures of all possible worlds in which f is true, i.e.

$$P(f) = \sum_{\omega \models f} m(\omega).$$

The probability of a formula f expresses the belief of our observer (agent) in the "truth" of the formula.

For example, consider the outcome (X_1, X_2) after tossing a pair of dice. We have 36 different possible worlds ω. in Ω. An observer (agent) who knows that one of the dice is falsified would have a certain opinion about the "true world". Another observer that has no prior knowledge at all about the dices would presumably consider all possible worlds being equal candidates to became the "true world", thus he will define his measure m by the obvious $m(\omega) = \dfrac{1}{36}$ for all $\omega \in \Omega$.

Consider the formula $X_1 + X_2 = 1$. There is no world ω in which this formula is true. Thus $P(X_1 + X_2 = 1) = 0$.

Consider the formula $X_1 + X_2 = 3$. There are two different worlds

$$\omega_1 \models (X_1 = 1) \wedge (X_2 = 2)$$

$$\omega_2 \models (X_1 = 2) \wedge (X_2 = 1)$$

in which the formula is true. Thus $P(X_1 + X_2 = 3) = \dfrac{1}{36} + \dfrac{1}{36} = \dfrac{1}{18}$.

The definition above supposes all the possible worlds are explored and this is not a practical approach. Let us give an alternative definition, without directly using measure function. Instead, we will accept four axioms (A1-A4) below.

(A1)   If the composed formula $f \Leftrightarrow g$ is a tautology (i.e. it is true in all worlds), then formulas f and g have the same probability:

$P(f) = P(g)$.

(This is obvious, both formulas f and g select the same possible worlds in which they are true.)

(A2) $P(f) \geq 0$ for any formula f.

(A3)   If t is a tautology, then $P(t) = 1$.

(A4)   If $\neg(f \wedge g)$ is a tautology, then

$P(f \vee g) = P(f) + P(g)$.

If we accept these intuitive axioms, then we can follow the model of (classical) Probability Theory. From these axioms, the following can be deduced.

Proposition 3.10. If f is a formula, then $P(f) + P(\neg f) = 1$.

Indeed, $f \vee \neg f$ is a tautology.

Proposition 3.11. If f and g are formulas, then

$$P(f) = P(f \wedge g) + P(f \wedge \neg g).$$

Indeed, $f \Leftrightarrow (f \wedge g) \vee (f \wedge \neg g)$ is a tautology.

Proposition 3.12. If X is a discrete random variable with finite range D and f is a formula, then

$$\sum_{x \in D} P(f \wedge X = x) = P(f).$$

In particular, if X is the Bernoulli distribution $Be(\pi)$, its range is $D = \{0, 1\}$; therefore

$$P(f \wedge X = 0) + P(f \wedge X = 1) = P(f).$$

Proposition 3.13. If f and g are formulas, then

$$P(f \vee g) = P(f) + P(g) - P(f \wedge g).$$

Indeed, $f \vee g \Leftrightarrow (f \wedge \neg g) \vee g$ is a tautology.

Given two formulas h and e, the **conditional probability** of h given e, denoted $p(h \mid e)$, is a measure of belief in a formula h based on the truth of another formula e.

Of course, the letter e stands for "evidence". It represents usually all of the agent's observation about the different worlds ω. And, of course, the letter h stands for "hypothesis".

$p(h \mid e)$ is said to be the **posterior probability** of the formula h. The usual probability $P(h)$ is the **prior probability** of h and is the same as $p(h \mid \text{true})$.

It is tempting to think "automatically" as follows: $p(h \mid e)$ means the probability of "h given e", which means the probability of "h if e", that means the probability of "$e \Rightarrow h$", that means the probability of "$(\neg e) \vee h$". Therefore $p(h \mid e) = P(\neg e \vee h)$. This conclusion is incorrect!

In fact, it can be shown that $P(e \Rightarrow h) = P(\neg e) + P(e) \cdot p(h \mid e)$.

Note the conditional probability $p(h \mid e)$ is very different from the probability of the implication $P(e \Rightarrow h)$. The latter is the common measure of all worlds for which h is true or e is false, i.e. is exactly $P(\neg e \vee h)$.

Suppose the evidence e is such that there exist worlds in which e is true. This evidence e induces a new measure

$$m_e : \Omega \to [0, 1]$$

for which all possible worlds ω such that e is false in ω have measure 0. To fulfill axiom (M2), the measures of the remaining worlds should be "normalized" so that the overall sum is 1 and not $P(e)$. An obvious solution is the following:

$$m_e(\omega) = \begin{cases} \dfrac{1}{P(e)} m(\omega) & \text{if } e \text{ is true in } \omega \\ 0 & \text{if } e \text{ is false in } \omega \end{cases}.$$

From here we obtain immediately

$$p(h \mid e) = \sum_{\omega \models h} m_e(\omega) = \frac{1}{P(e)} \sum_{\omega \models h \wedge e} m(\omega)$$

thus the well-known relation

$$p(h \mid e) = \frac{P(h \wedge e)}{P(e)}$$

or, in other terms,

$$P(h \wedge e) = P(e) \cdot p(h \mid e).$$

The following generalization (known as the "chain rule"), valid for a multiple conjunction $f_1 \wedge f_2 \wedge f_3 \wedge \ldots \wedge f_n$ of formulas, is obvious

$$P(f_1 \wedge f_2 \wedge f_3 \wedge \ldots \wedge f_n) = P(f_1) \cdot p(f_2 \mid f_1) \cdot p(f_3 \mid f_1 \wedge f_2) \cdot \ldots$$
$$\ldots \cdot p(f_n \mid f_1 \wedge f_2 \wedge f_3 \wedge \ldots \wedge f_{n-1}).$$

When using probabilities, there are two ways to state that a formula f "is true":

- either to write $P(f) = 1$ (this means that f is true in all worlds),
- or to condition on f, that means to use f as a component of the evidence e in a conditional expression $p(h \mid e)$. However, in doing this we restrict ourselves only to the worlds in which f is true, all the others are neglected.

When reasoning under uncertainty, a rational agent is updating his beliefs in the light of new evidence. Suppose he possess knowledge k and "knows" the posterior probability $p(h \mid k)$ of the hypothesis h. Later he detects a new piece of evidence e, which should be taken into account.

Of course, the new piece of evidence e is "added" to the old knowledge k, giving the new knowledge $k \wedge e$. The old posterior probability $p(h \mid k)$ becomes now the prior probability, and a new posterior probability has to be calculated. The Bayes rule is used to obtain:

$$p(h \mid k \wedge e) = p(h \mid k) \cdot \frac{p(e \mid h \wedge k)}{p(e \mid k)}.$$

Thus, in order to update the posterior probability, the agent should possess prior "estimates" of the conditional probabilities $p(e \mid h \wedge k)$ and $p(e \mid k)$. Usually the first is easier to obtain. As for the second, the following result could be used:

If $\{h_1, h_2, ..., h_n\}$ is an exclusive and exhaustive finite family of statements (i.e. a covering set), representing all possible hypotheses, then

$$p(e \mid k) = \sum_i p(e \mid h_i \wedge k) \cdot p(h_i \mid k).$$

Let us recall the Bayes' theorem in its most general expression:

$$p(h_i \mid e) = \frac{p(e \mid h_i) \cdot P(h_i)}{\sum_{j=1}^{n} p(e \mid h_j) \cdot P(h_j)}$$

where $\{h_1, h_2, ..., h_n\}$ are mutually exclusive and exhaustive hypotheses, and e is the "evidence".

Example [Degoulet and Fieschi 1999]. The observer-agent is a physician, and he knows from his experience that in most cases a pain in the right lower quadrant is caused by appendicitis, let us say he estimates

p(pain in right lower quadrant | patient has appendicitis) = 0.8.
    He knows that there is a small probability that appendicitis cause also pain in the left lower quadrant, let us say he estimates
p(pain in left lower quadrant | patient has appendicitis) = 0.1.
    Salpingitis is a disease possibly causing pain in the left lower quadrant for women; our physician estimates
p(pain in left lower quadrant | patient has salpingitis) = 0.5.
    He knows also the probability for a new admitted female patient in the hospital to be diagnosed with appendicitis res. salpingitis
P(new female patient admitted has appendicitis) = 0.08,
P(new female patient admitted has salpingitis) = 0.05.
    (How? Suppose he is thinking this way: last year we had 1000 female patients in the hospital and 80 had appendicitis surgery. Therefore, ...)

Let us put this knowledge in a table:

Possible disease D	Prior probability $P(D)$	p(pain in right quadrant $\mid D$)	p(pain in left quadrant $\mid D$)
Appendicitis	0.08	0.80	0.10
Salpingitis	0.05	0.50	0.50
Any other	0.87	0.05	0.05

The Bayes' Theorem allows us to calculate posterior probabilities:
p(female patient has appendicitis | pain in right lower quadrant) =

$$= \frac{0.80 \cdot 0.08}{0.80 \cdot 0.08 + 0.50 \cdot 0.05 + 0.05 \cdot 0.87} = 0.483 \, ,$$

p(female patient has salpingitis | pain in right lower quadrant) =

$$= \frac{0.50 \cdot 0.05}{0.80 \cdot 0.08 + 0.50 \cdot 0.05 + 0.05 \cdot 0.87} = 0.189$$

therefore "appendicitis" is the first diagnostic to be considered when a woman is complaining of pain in the right lower quadrant.

What happens if the woman is complaining of pain in both lower quadrants? In general – from the theoretical point of view – it is difficult to give an answer.

What if we suppose the evidences $e_1 =$ "pain in left lower quadrant" and $e_2 =$ "pain in right lower quadrant" are independent? (Let us say we have reasons to accept this as a fact.) Now the Bayes' Theorem becomes:

$$p(h_i \mid e_1 \wedge e_2) = \frac{p(e_1 \wedge e_2 \mid h_i) \cdot P(h_i)}{\sum_{j=1}^{n} p(e_1 \wedge e_2 \mid h_j) \cdot P(h_j)}$$

and independence condition between evidences is expressed in the following way

$$p(e_1 \wedge e_2 \mid h_i) = p(e_1 \mid h_i) \cdot p(e_2 \mid h_i) \, .$$

Thus:

p(appendicitis | pain in both lower quadrants) =

$$= \frac{0.80 \cdot 0.10 \cdot 0.08}{0.80 \cdot 0.10 \cdot 0.08 + 0.50 \cdot 0.50 \cdot 0.05 + 0.05 \cdot 0.05 \cdot 0.87} = 0.306 \text{ (decreases)}.$$

On the other hand,

p(salpingitis | pain in both lower quadrants) = 0.552

raises high enough to put "salpingitis" as the first diagnostic to be considered.

In general, let us consider the following equality

$$p(h \mid k \wedge e) = p(h \mid k) \, .$$

This equality expresses a very simple idea: the new piece of evidence e is completely irrelevant. We can expect irrelevant data received by an agent to overwhelm the relevant one. The treatment of such situation will be exposed in the context of Bayesian networks (Chapter 5).

Consider a finite set of statements, where some are evidences (observations) and others hypothesis. Bayes' Theorem, in a general form, concerns observations $e_1, e_2, ..., e_m$ that influence hypotheses $h_1, h_2, ..., h_n$ (supposed mutually exclusive and exhaustive). The formula is:

$$p(h_i \mid e_1 \wedge e_2 \wedge \dots \wedge e_m) = \frac{p(e_1 \wedge e_2 \wedge \dots \wedge e_m \mid h_i) \cdot P(h_i)}{\displaystyle\sum_{j=1}^{n} p(e_1 \wedge e_2 \wedge \dots \wedge e_m \mid h_j) \cdot P(h_j)}.$$

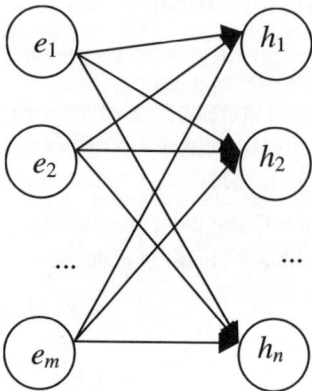

Fig. 3.17. Diagram of influences

The formula above supposes that it is possible to estimate all conditional probabilities for all possible combinations of e_k for all possible hypothesis h_i (thus 2^m conditional probabilities for each hypotheses) and also to estimate the a priori probabilities of h_i.

The effort is huge. A possible solution is to accept the conditional independence of observations regarding the hypotheses:

$$p(e_k \wedge e_l \mid h_i) = p(e_k \mid h_i) \cdot p(e_l \mid h_i).$$

The number of conditional probabilities that have to be estimated is now m and not 2^m (for each hypothesis). The detailed treatment will be approached in the context of Bayesian networks.

3.18 Solved Exercises

1) Consider a sigma-algebra $\mathcal{E}$ of events over the universe Ω and a function $Q : \mathcal{E} \rightarrow [0,1]$ satisfying the conditions

(Π1) $Q(\Omega) = 1$,

(Π3)   If $E, E_1, E_2, \dots$ are   events   such   that $E_1 \subseteq E_2 \subseteq E_3 \subseteq \dots$ and

$\displaystyle\bigcup_{n=1}^{\infty} E_n = E$, then $\displaystyle\lim_{n\to\infty} Q(E_n) = Q(E)$

Show that Q satisfies (Π2), i.e. is a probability.

2) If you have only a fair coin available, how could you produce an event of probability $\frac{1}{3}$?

3) Consider all the families having exactly two children. If $P(\text{girl}) = \pi$, show the two conditional probabilities $p(\text{two girls} \mid \text{elder child a girl})$ and $p(\text{two girls} \mid \text{at least one girl})$ are different.

4) Last year Peter decided to buy a new computer. The disk of this computer failed yesterday, after working continuously for ten months. Peter found the following statistical data this morning in a computer magazine:

Disks make	Weight on the market	Reliability of products
Alphadisk	35%	20
Betadur	30%	15
Gammamix	15%	15
Others	20%	5

(The reliability is expressed by the inverse of the probability of a failure during the first year of operation.)

Estimate for Peter, before opening the system unit, the chances that "Alphadisk" is the make.

5) In a small mining town, 212 inhabitants (from the total of 1018) exhibit a form of silicosis disease. The men are most touched, because 180 men are ill. However, the other 412 men are not ill.

Find out the probability that the woman just entering in the medical cabinet has this silicosis disease.

6) Denote $p_k = P(b(n, \pi) = k)$ for $k \in \{0, 1, 2, ..., n\}$. Show that the values p_k have a single maximum, and this lies between $np + p - 2$ and $np + p - 1$.

7) Let X be a random variable defined on $(\Omega, \mathcal{A})$. Prove that $\{\omega \in \Omega \mid X(\omega) = a\} \in \mathcal{A}$ for any $a \in \mathbf{R}$.

8) Suppose the random variable X can only assume integer values 0, 1, 2, ... Show that $E(X) = \sum_{n=0}^{\infty} P(X > n)$.

9) Assume that a specialist physician will diagnose correctly a patient with probability 0.95. This means that:

a) Given an ill patient, the probability is 0.95 that the physician will diagnose him with the correct illness, and

b) Given a healthy patient, the probability that the physician will declare him as such is also 0.95.

Suppose the family caring physician is very good, such that 99% of the people that he sends to the specialist are actually ill. Compute:

a) The probability that a person is healthy, given that the specialist declares him as healthy, and

b) The probability that a person is healthy, given that the specialist declares him as ill.

10) Suppose we have three propositional variables X, Y, Z, i.e. random variables of Bernoulli type

$$X = Be(\xi): \begin{pmatrix} 0 & 1 \\ 1-\xi & \xi \end{pmatrix}, \quad Y = Be(\eta): \begin{pmatrix} 0 & 1 \\ 1-\eta & \eta \end{pmatrix}, \quad Z = Be(\zeta): \begin{pmatrix} 0 & 1 \\ 1-\zeta & \zeta \end{pmatrix}$$

where 1 res. 0 are assimilated with true, res. false. Denote by x res. $\neg x$ the proposition $X = 1$ res. $X = 0$ (analogously for variables Y and Z).

Suppose we know all the eight joint probabilities:

$$P(x \wedge y \wedge z) = 0.3 \qquad P(x \wedge \neg y \wedge z) = 0.2$$
$$P(x \wedge y \wedge \neg z) = 0.2 \qquad P(x \wedge \neg y \wedge \neg z) = 0.1$$
$$P(\neg x \wedge y \wedge z) = 0.05 \qquad P(\neg x \wedge \neg y \wedge z) = 0.05$$
$$P(\neg x \wedge y \wedge \neg z) = 0.1 \qquad P(\neg x \wedge \neg y \wedge \neg z) = 0$$

If we are given $\neg z$ as evidence, calculate the conditional probability $p(y \mid \neg z)$. If we are given x and $\neg z$ as evidence, calculate the conditional probability $p(y \mid x \wedge \neg z)$.

11) f and g are formulas. Prove that $P(\neg(f \Leftrightarrow g)) \le P(f) + P(g)$.

12) Given $P(f) = \alpha$, $p(f \mid g) = \beta$ and $p(f \mid \neg g) = \gamma$, compute the probability $P(f \Rightarrow g)$.

Solutions. 1) Indeed, let $A_1, A_2, ..., A_n, ...$ be a sequence of events such that $A_i \cap A_j = \varnothing$ for $i \ne j$. Denote $E_n = \bigcup_{k=1}^{n} A_k$. It is immediate that $A_{n+1} = E_{n+1} - E_n$. Moreover, $\bigcup_{n=1}^{\infty} E_n = \bigcup_{n=1}^{\infty} A_n$ and $Q(E_n) = \sum_{k=1}^{n} Q(A_k)$, thus $\lim_{n} Q(E_n) = \sum_{k=1}^{\infty} Q(A_k)$. The conclusion $Q(\bigcup A_n) = \sum Q(A_n)$ is obvious.

2) A "fair coin" means that, after one toss, the probability of a head (H) and the probability of a tail (T) both equal $\frac{1}{2}$. Of course, the Bernoulli distribution

$$Be(\frac{1}{2}) = \begin{pmatrix} 0 & 1 \\ \frac{1}{2} & \frac{1}{2} \end{pmatrix}$$

represents the number of heads obtained, after one toss. By

repeating the toss n times the number of heads obtained is represented by the binomial distribution $b(n, \frac{1}{2})$. Any event related to n tosses is expressed in terms of these:

$b(n, \frac{1}{2}) = k$, where $k \in \{0, 1, ..., n\}$

(more precisely as disjunctions) and we know that

$$P(b(n, \tfrac{1}{2}) = k) = \binom{n}{k} \cdot (\tfrac{1}{2})^n = \frac{\binom{n}{k}}{2^n}.$$

Because $\dfrac{1}{3}$ cannot be expressed as a dyadic number $\dfrac{m}{2^n}$, it is rather obvious that, using only our fair coin, no event having this probability can be ever produced.

However, the latter sentence does not refer to "conditional events"! Indeed, take $n = 4$ and denote by A the event $b(4, \frac{1}{2}) \leq 3$, and by B the event $b(4, \frac{1}{2}) \leq 1$. Then $P(A) = \dfrac{15}{16}$, $P(B) = \dfrac{5}{16}$, and $B \cap A = B$. Hence

$$p(B|A) = \frac{P(B \cap A)}{P(A)} = \frac{1}{3} \ !$$

3) It is a good idea to build a tree of "events"

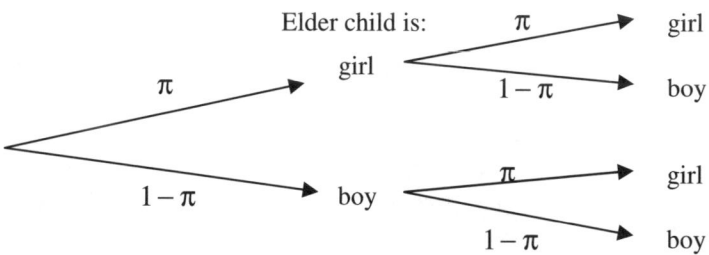

It is clear that $p(\text{two girls} \mid \text{elder child a girl}) = \pi^2$. On the other hand, $p(\text{two girls} \mid \text{at least one girl}) = \pi + \pi \cdot (1 - \pi) = 2\pi - \pi^2$.

4) We know, from statistical data presented in the magazine, that
$P(Alphadisk) = 0.35$, $P(Betadur) = 0.30$,
$P(Gammamix) = 0.15$, $P(Others) = 0.20$ and

$p(failure \mid Alphadisk) = \dfrac{1}{20}$, $p(failure \mid Betadur) = \dfrac{1}{15}$,

$p(failure \mid Gammamix) = \dfrac{1}{15}$, $p(failure \mid Others) = \dfrac{1}{5}$.

Our task is to estimate $p(Alphadisk \mid failure)$. By using Theorem III.1, we obtain $p(Alphadisk \mid failure) = 0.18$. Thus, there are 18% chances to blame Alphadisk.

5) The known data are placed in the table below:

	Ill	Not ill	Total
Men	180	412	
Women			
Total	212		1018

It is very easy to find out that 32 women are ill, from a total of 426. The incidence of silicosis in women is $\dfrac{32}{426} = 0.075117$, hence we estimate our probability at 7.5%.

6) Indeed, $p_k = \dfrac{n!}{k!\cdot(n-k)!} \cdot \pi^k \cdot (1-\pi)^{n-k}$. If we compare two consecutive values, p_k and p_{k+1}, we obtain $\dfrac{p_{k+1}}{p_k} = \dfrac{n-k}{k+1} \cdot \dfrac{\pi}{1-\pi}$. The values p_k increase until the maximum, then decrease. The maximum p_r is found from $p_r \le p_{r+1} > p_{r+2}$.

7) The sigma-algebra of Borel sets is generated by intervals $[a,b)$ The relation

$$\{a\} = [a, a+1) - \bigcup_{n=1}^{\infty} \left[a + \frac{1}{n+1}, a + \frac{1}{n} \right) \text{ shows that } \{a\} \text{ is a Borel set, hence}$$

$X^{-1}(\{a\}) \in \mathcal{A}$.

8) The expectation is $E(X) = \displaystyle\sum_{n=0}^{\infty} n \cdot P(X = n)$. On the other hand, it is obvious that $P(X = n) = P(X > n) - P(X > n+1)$.

9) Consider the following events: I = "the person sent to the specialist by the family caring physician is ill", J = "the specialist diagnoses a person with the correct illness". We are given:

$p(J \mid I) = 0.95$, $p(\neg J \mid \neg I) = 0.95$, $P(I) = 0.99$, $P(\neg I) = 0.01$.

We want a) $p(\neg I \mid \neg J)$ and b) $p(\neg I \mid J)$. These conditional probabilities are found from Bayes' theorem:

$$p(\neg I \mid \neg J) = \frac{p(\neg J \mid \neg I) \cdot P(\neg I)}{p(\neg J \mid \neg I) \cdot P(\neg I) + p(\neg J \mid I) \cdot P(I)} = \ldots = 0.1610,$$

$$p(\neg I \mid J) = \frac{p(J \mid \neg I) \cdot P(\neg I)}{p(J \mid \neg I) \cdot P(\neg I) + p(J \mid I) \cdot P(I)} = 0.616.$$

10) Of course, we could make appeal to the formula

$$p(y \mid \neg z) = \frac{P(y \wedge \neg z)}{P(\neg z)}.$$

in which the two probabilities, yet unknown, $P(y \wedge \neg z)$ and $P(\neg z)$, appear. To obtain the first one we use (see Proposition 3.11)

$$P(y \wedge \neg z) = P(x \wedge y \wedge \neg z) + P(\neg x \wedge y \wedge \neg z) = 0.2 + 0.1 = 0.3.$$

However, we need the other, i.e. $P(\neg z)$. A solution is to calculate the ingredients needed in the analog formula $p(\neg y \mid \neg z) = \dfrac{P(\neg y \wedge \neg z)}{P(\neg z)}$ and to use Proposition 3.3:

$p(y \mid \neg z) + p(\neg y \mid \neg z) = 1$.

Since $P(\neg y \wedge \neg z) = P(x \wedge \neg y \wedge \neg z) + P(\neg x \wedge \neg y \wedge \neg z) = 0.1 + 0 = 0.1$, it follows $\dfrac{0.3}{P(\neg z)} + \dfrac{0.1}{P(\neg z)} = 1$, therefore $P(\neg z) = 0.4$ and from here the result we need: $p(y \mid \neg z) = \dfrac{0.3}{0.4} = 0.75$. The conditional probability $p(y \mid x \wedge \neg z)$ is obtained directly as $P(y \wedge x \wedge \neg z)/P(x \wedge \neg z)$.

11) Indeed, $\neg(f \Leftrightarrow g)$ is equivalent to $(\neg f \wedge g) \vee (f \wedge \neg g)$. If we denote $P(f \wedge g) = \alpha$, $P(f \wedge \neg g) = \beta$, $P(\neg f \wedge g) = \gamma$, $P(\neg f \wedge \neg g) = \delta$, then $P(\neg(f \Leftrightarrow g)) = \beta + \gamma$. On the other hand, $P(f) = P(f \wedge g) + P(f \wedge \neg g) = \alpha + \beta$, $P(g) = \alpha + \gamma$. The inequality is obvious, and it turns into equality only if $\alpha = 0$.

12) The formula $f \Rightarrow g$ can be expressed as $\neg f \vee g$, or as $\neg(f \wedge \neg g)$. Hence $P(f \Rightarrow g) = 1 - P(f \wedge \neg g) = 1 - p(f \mid \neg g) \cdot P(\neg g)$. On the other hand, $P(f) = p(f \mid g) \cdot P(g) + p(f \mid \neg g) \cdot P(\neg g)$. Thus $P(f \Rightarrow g) = 1 - \gamma \cdot P(\neg g)$ and $\alpha = \beta + (\gamma - \beta) \cdot P(\neg g)$. Of course α, β and γ cannot be arbitrary.

4 Statistical Inference

4.1 Inferring Scientific Truth: Tests of Significance

Human knowledge advances continuously; scientific researchers gain new knowledge every day. What methods they use? When a phenomenon occurs, reasonable people try to detect what causes it, and put forward hypotheses that seem plausible. By observing further occurrences, some hypotheses are enhanced, some others diminished or even rejected, i.e. the plausibility of each explanatory hypothesis is re-evaluated.

Significance testing is a particular method to assess plausibility degrees. Its particularity is clear: it refers to special kinds of hypotheses called statistical hypotheses.

From a common sense approach, testing a particular assumption (i.e. a *hypothesis*) that our personal experience tells us to believe in it is easy to explain: we assume the hypothesis is true, and then we compare observations (i.e. data obtained from the real world) with logical consequences of our hypothesis. If the available observations are compatible with the expected consequences, then we continue to believe – and in most cases we strengthen our belief – in our assumption. Of course, if what is observed does not fit close enough to what we expect, then our belief in the assumption will diminish, sometimes we may reject entirely our hypothesis.

(Notice some fuzziness in the previous paragraph: the precise meaning of "close enough" is left to anyone of us to decide for himself. And to bear any unpleasant consequence a wrong decision would imply!)

Obviously, hypotheses put forward by scientific researchers are known as scientific hypotheses.

Conducting a significance test (known also as hypothesis testing) is a method employed to test a believed assumption about an entire population, using data gathered from a sample. In general, the result of a significance test is expressed as a number. This number reflects how likely the obtained value of a descriptive statistics – which is computed using the data from that sample – may have come from a random sample.

The approach of Ronald A. Fisher[1] was dedicated to scientific researchers: the validity of a scientific hypothesis is established on the basis of a single test, with the option of suspending judgment when the results are not "clear enough". Two options are available in this approach: either to "reject the null hypothesis", or to suspend judgment (there is not enough data to conclude). However, physicians are seldom "research workers"; on the contrary, most of their work resembles that of decision

---

[1] Ronald Aylmer Fisher (1890-1962), English geneticist and statistician.

E. Roventa and T. Spircu: Management of Knowledge Imperfection, STUDFUZZ 227, pp. 89–131.
springerlink.com   © Springer-Verlag Berlin Heidelberg 2009

makers. For a decision maker a decision must be taken on the basis of limited information. A rational decision maker tries to minimize the cost of a wrong decision. The approach he/she follows when confronted with two competing hypotheses is clear: a choice must be made; the decision is taken on the basis of information previously gathered from samples.

Either as a scientific researcher, or as a decision maker, You are in position to make a rational decision – after conducting tests of significance – only when you thoroughly understand the idea of these tests. This involves two aspects:

1) On one hand, you should understand those questions for which tests of significance provide (at least partial) answers, and

2) On the other hand, you have to acquire a sense of the nature of the information these tests provide.

From the point of view of understanding the world and from the logical point of view, Fisher's approach is easier to explain: a scientific hypothesis refers to theoretical populations, which usually have an infinite number of individuals and which are represented by continuous distributions. This scientific hypothesis is replaced by a statistical hypothesis, which is expressed by means of a parameter of that population (such as a proportion, an average etc.). The value of the parameter is estimated by exploiting data extracted from a sample, and then compared to some expected value. The discrepancy between the two will influence our belief in the (validity of the) scientific hypothesis.

A statistical hypothesis that is associated to a scientific hypothesis is based, therefore, on a "small" sample extracted from a finite (possibly "large") population. A first source of error has its origins in the identification of the scientific hypothesis with the associated statistical one. However, when using statistical methods, in fact we do identify these two hypotheses and we try to evaluate the risk of error.

Scientific researchers use on a large scale a fallacious reasoning, called in Latin *abductio*:

$$\frac{H \Rightarrow O \ , \ O}{H}$$

and a correct one, called in Latin *modus tollens*:

$$\frac{H \Rightarrow O \ , \ \neg O}{\neg H}.$$

Here the letter H represents a scientific hypothesis, and the letter O represents an observation. In both arguments above the implication $H \Rightarrow O$ is considered an acquired knowledge, i.e. is accepted as "absolutely certain". Of course, observing O increases our belief in the hypothesis H (however, it cannot assure us that H is "valid", nor "true"), and observing $\neg O$ excludes H as a valid hypothesis. Thus, in classical logic we cannot prove a hypothesis (to be true) but we can disprove a hypothesis. From this "classical" point of view, a scientific truth is a statement that has a very low probability of being proven incorrect in the future (see [Popper 1959]).

In short, the idea of hypothesis testing (i.e. tests of significance) is simple: our statistical hypothesis will serve as an alternative to another hypothesis – the so-called "null hypothesis" – that is raised only to be rejected. By accepting the validity ("truth") of the null hypothesis, some statistical consequences will follow, and these will be confronted with the observed data. Any strong evidence against the null hypothesis will serve as a justification in favor of the alternative.

4.2 Relation "Alternative Hypothesis – Null Hypothesis"

As stated above, a statistical hypothesis is a statement about a population parameter (or several population parameters). Such a statement is related to (or is a logical consequence of) a corresponding "scientific hypothesis".

Let us present, by some examples, how the relation between the two kinds of hypotheses appears. Consider the following statements:

(1) At the age of 10, girls are more intelligent than boys,
(2) "Very old" age is a significant predictor of Alzheimer,
(3) Children are more creative than adults,
(4) Drug A helps recover people better than drug B,
(5) Male and female physicians earn different salaries,
(6) Patients recover after a standard treatment,
(7) People following a weekly diet prescribed by the famous Dr. C will loose exactly 2 kg,
(8) Drug D has no effect on tuberculosis,
(9) The effects of drug E on male and female patients are similar.

Professionals may declare, as personal beliefs, perhaps as a result of their (possibly long) personal experience, any one of these nine statements.

However, there is a clear distinction between the last three and the first six: the last three express equality, similarity or coincidence (notice that "has no effect" means "does not change the situation", or "the situation after a drug treatment is the same as the situation before"). On the contrary, the first six statements express inequality, dissimilarity, or difference.

This distinction is essential for hypothesis testing. It is essential to point out that hypothesis testing can be used to confirm only scientific hypotheses that are expressed as inequalities, dissimilarities, or differences; in no way equalities like that expressed in (7) may be found "true" by hypothesis testing. Perhaps what our professional (could it be Dr. C himself?) has had in mind has been this:

(7') People following a weekly diet prescribed by the famous Dr. C will loose at least 2 kg

and in this form it may serve as a start to a statistical hypothesis testing.

Let us replace our seven scientific hypotheses (1)-(6) and (7') above by their corresponding statistical hypotheses. Of course, some parameters of the respective populations, such as averages, proportions … will be involved:

(1_a) The average IQ of age 10 girls is bigger than the average IQ of age 10 boys,
(2_a) The incidence of Alzheimer disease is higher in "very old" people (compared to "old" people),
(3_a) The average creativity index of children is higher than that of adults,

(4_a) The proportion of recovered people is larger in those treated with drug A (compared to those treated with drug B),

(5_a) The average salary of male physicians is higher than the average salary of female physicians,

(6_a) The average health status of patients, after a standard treatment, is better than before the treatment,

(7_a) The average weekly loss of weight of people that follow the diet prescribed by the famous Dr. C is at least 2 kg.

All these sentences will serve as alternative hypotheses in tests of significance. In general, in a significance test, the alternative hypothesis is a statement about a population parameter, which replaces the (assumed plausible) scientific hypothesis. (Notice that in all examples before averages or proportions were involved as population parameters.)

It is customary to call alternative hypothesis and to denote by H_a (or H_1) the scientific hypothesis under question, considered as an assessment expressing an inequality, dissimilarity, or a difference.

From a logical point of view, we may declare another assessment in the same terms, this time expressing equality or the reverse inequality, similarity, or coincidence. The latter assumption is denoted by H_0 and is called the null hypothesis. Following R. A. Fisher, the null hypothesis is raised – as a complement to the alternative hypothesis – only to be rejected, and by rejecting it we automatically accept as "true" our initial scientific hypothesis.

Let us present such assessments for the seven examples above:

(1_0) The average IQ of age 10 girls is equal to the average IQ of age 10 boys,

(2_0) The incidence of Alzheimer disease among "very old" people is the same as among "old" people,

(3_0) The average creativity index of children is less than that of adults,

(4_0) The proportions of recovered people after treatment with drugs A res. B are the same,

(5_0) The average salary of male physicians coincides with the average salary of female physicians,

(6_0) The average health status of patients, after a standard treatment, is not suffering any change,

(7_0) The average weekly loss of weight of people that follow the diet prescribed by the famous Dr. C is exactly 2 kg.

R. A. Fisher gave the name "null hypothesis" because this hypothesis should be "nullified". The name was retained and survived probably because in most cases the null hypothesis could be written as equality "with null":

$$(H_0) \quad f(\pi) = 0$$

where f is a function of the parameters π of the populations involved in the test. Perhaps the best example is:

$$(1_0) \quad \mu_g - \mu_b = 0$$

where the parameters μ_g and μ_b represent the average IQ of age 10 girls res. the average IQ of age 10 boys.

There is always possible that the null hypothesis is true, hence rejecting it may lead to an error. The probability of such an error is known as the p-value and is usually interpreted as the risk of accepting our scientific hypothesis as being true.

Supposing some information about the distribution of the population is available, the only source of errors appears to be the manner in which the individuals from the sample are selected. When the individuals are randomly selected (it is said that the sample is randomly selected), the differences between the observed outcome and the expected outcome are explained only by chance factors. We could set a threshold on these differences, which in our intent will separate "small" (acceptable) differences, from "large" (unacceptable by chance) differences. This threshold is in fact determined by the significance level.

4.3 Hypothesis Testing, the Classical Approach

As we pointed out above, in any hypothesis testing data obtained only from a sample are considered and processed. Of course, the sampling procedure is supposedly random, and usually it is accepted that the populations under study are distributed normally.

In its classical approach, which is decision-oriented, a hypothesis testing involves five consecutive steps, as follows:

Step 1: Specify the alternative hypothesis, then the null hypothesis.

Step 2: Choose the statistic that is adapted to the concrete situation. (The word *statistic* means "formula involving data extracted from a sample".)

Step 3: Choose the significance level, and then determine the associate threshold.

Step 4: Compute the value of the statistic, using data obtained from the (randomly selected) sample.

Step 5: Decide, by comparing the computed value to the threshold determined by the significance level, whether or not to reject the null hypothesis.

The discussion around hypothesis testing begins with the last step. Here, a decision maker has to decide either to reject H_0 (and, therefore, to accept the alternative hypothesis H_a as "valid" or "true"), or not to reject the H_0. In reality, H_0 is either true or false – however, our decision maker totally ignores the real status of the world. The four different possibilities resulting from this situation are as follows:

		Reality (unknown)	
		H_0 is false	H_0 is true
Decision	Reject H_0	Correct!	**Erroneous (type I error)**
	Do not reject H_0	Erroneous (type II error)	Correct!

In two of them the decision is correct. However, when rejecting a true H_0 our decision maker commits a type I error. Also, when failing to reject a false H_0 he/she commits a type II error.

Of maximum importance in hypothesis testing is the type I error. Its probability, i.e. the number

$$\alpha = p(\text{erroneous decision} \mid H_0 \text{ is true})$$

is the significance level whose value was previously chosen (in Step 3).

Of course, any decision maker wants to keep the significance level as low as possible – in fact it is the probability of an error! Thus, values such as $\alpha = 0.05$ are common, and in medical sciences even values as small as $\alpha = 0.001$ are recommended.

4.4 Examples: Comparing Means

Let us present some typical examples of significance testing.

Example 1. Suppose the alternative hypothesis is (7_a) above, and the null hypothesis is (7_0) above.

(Notice the alternative hypothesis is expressed as "at least", i.e. is one-tailed.)

We start from the assumption that (7_0) is true, i.e. it is true that a person following the diet prescribed by the famous Dr. C will lose weight, on average, 2 kg each week. Implicitly, we suppose that the weekly weight loss is a random variable, normally distributed with mean $\mu = 2$ and variance σ^2 (unknown). A sample of volume N, extracted from the population of individuals following the diet, is in fact a sequence $X_1, X_2, ..., X_N$ of the corresponding weekly weight losses, which in fact are independent random variables of type $N(\mu, \sigma^2)$.

It is well known that, under these circumstances, the sample mean,

$$M = \frac{1}{N}(X_1 + X_2 + ... + X_N),$$

as a random variable, is again a normal distribution, with the same mean μ as any X_i, and variance $\dfrac{\sigma^2}{N}$. Hence the population of values $\dfrac{M - 2}{\sigma / \sqrt{N}}$ obtained from samples of volume N can be considered a random variable of type $N(0, 1)$, i.e. a standard normal distribution.

However, as we pointed out above, σ^2 is unknown. Usually it is estimated by the so-called sample variance

$$S^2 = \frac{1}{N-1}((X_1 - M)^2 + (X_2 - M)^2 + ... + (X_N - M)^2)$$

and we are interested in the formula

$$T = \frac{M - 2}{S / \sqrt{N}}.$$

This formula will be chosen (in Step 2) as the statistic to be used in Step 4.

It is well known in Statistics that T is a random variable of Student type; more precisely its type is $t(N-1)$.

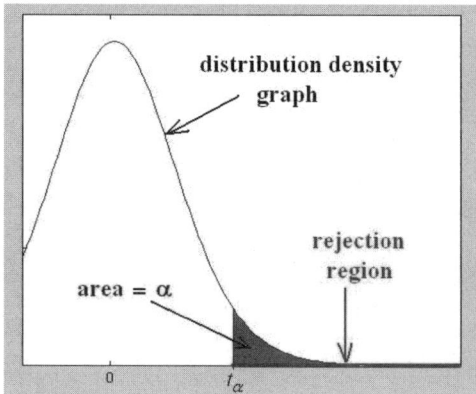

distribution density graph

rejection region

area = α

0   t_α

Fig. 4.1. Rejection region in a unilateral test

Most of the values of T are concentrated around 0. However, because values $M < 2$ will not support the one-tailed hypothesis (7ₐ), we are interested only in positive values of T. The positive values outside the interval $[0, t]$ – which is determined by a critical value t – are considered as significantly different from 0. These values are from the so-called rejection region, because in case of apparition of such a value our decision maker will reject the null hypothesis (see the Figure 4.1 above).

Once the significance level α has been chosen (in Step 3), the threshold (i.e. the critical value) $t_\alpha > 0$ that delimits the rejection region $(t_\alpha, +\infty)$ is uniquely determined (and computationally well approximated) from the condition $P(T > t_\alpha) = \alpha$, which is in fact equivalent to

$$P(T \leq t_\alpha) = 1 - \alpha.$$

Hence, the hypothesis testing goes as follows: after choosing a convenient significance level α, compute immediately the threshold $t_\alpha > 0$ from the condition

$$\Theta(t_\alpha) = 1 - \alpha$$

where Θ is the distribution function of the random variable $t(N-1)$. Select then a random sample of volume N, obtain data $x_1, x_2, ..., x_N$ from the sampled individuals, and compute the value

$$t = \frac{m - 2}{s / \sqrt{N}}$$

where $m = \dfrac{1}{N}(x_1 + x_2 + \ldots + x_N)$ and

$$s = \sqrt{\frac{1}{N-1}\left((x_1 - m)^2 + (x_2 - m)^2 + \ldots + (x_N - m)^2\right)} .$$

The final decision in Step 5 is taken considering only the relation between this computed value t and t_α. Namely, if $t > t_\alpha$, reject the null hypothesis.

As a particular case, let $N = 10$ and suppose the significance level $\alpha = 0.05$ is chosen. From here – looking into a table of the $t(9)$ distribution, or using a special function such as TINV in *Microsoft Excel* – we obtain the threshold $t_{0.05} \approx 2.2622$. Suppose the following data are obtained from the ten individuals of the sample:

Individual	Weekly weight loss	Individual	Weekly weight loss
1	2.3 kg	6	2.2 kg
2	2.8 kg	7	2.2 kg
3	2.1 kg	8	2.6 kg
4	3.0 kg	9	2.4 kg
5	2.3 kg	10	2.1 kg

Notice for all individuals in the sample the weekly weight loss is greater than 2 kg. This fact itself strongly supports the alternative! The sample mean is

$$\frac{1}{10}(2.3 + 2.8 + 2.1 + 3.0 + 2.3 + 2.2 + 2.2 + 2.6 + 2.4 + 2.1) = 2.4 \ (\text{kg})$$

and the sample variance is $s^2 \approx 0.0933$ ($s \approx 0.3055$). Therefore,

$$t \approx \frac{2.4 - 2}{0.3055 / \sqrt{10}} \approx 4.1404 .$$

Notice $t > t_{0.05}$; hence we are entitled to reject the null hypothesis, thus to accept as true the alternative hypothesis (7_a).

Suppose the significance level is lowered to $\alpha = 0.005$. Now the threshold is $t_{0.005} \approx 3.6896$, and we still have $t > t_{0.005}$. Even with this significance level $\alpha = 0.005$ (ten times less as before), we reject the null hypothesis and we consider the alternative hypothesis (7_a) to be true. However, if we lower again the significance level, this time to $\alpha = 0.001$, the new threshold will be $t_{0.001} \approx 4.7809$ and the computed value $t \approx 4.1404$ is no longer in the rejection region. We fail to reject the null hypothesis!

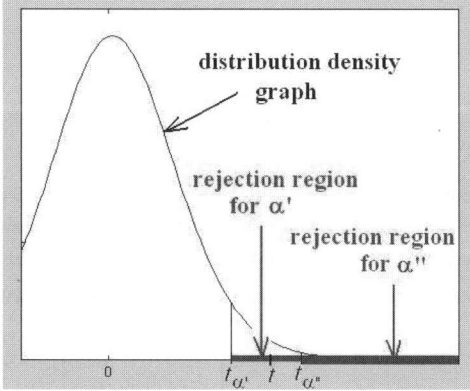

Fig. 4.2. Modifying the rejection region

Imagine a continuous change of the significance level from $\alpha' = 0.005$ (for which $t > t_{\alpha'}$, i.e. t is in the rejection region for α') to $\alpha'' = 0.001$ (for which $t < t_{\alpha''}$, i.e. t is not in the rejection region for α'' – see Figure 4.2).. There exists a particular significance level α^*, between α' and α'', such that t will be exactly the critical value t_{α^*}. This particular significance level is known as the p-value of the alternative hypothesis.

Its interpretation is clear: it is the smallest significance level that allows us to accept the alternative hypothesis as true – by rejecting the null hypothesis –, based on the data from the chosen sample only. Many people interpret this p-value as the risk of accepting as true the alternative hypothesis (based on the given sample).

Let us consider, as another particular case, a second sample:

Individual	Weekly weight loss	Individual	Weekly weight loss
1	1.6 kg	6	1.6 kg
2	2.8 kg	7	1.7 kg
3	1.6 kg	8	2.6 kg
4	3.0 kg	9	2.4 kg
5	1.9 kg	10	1.8 kg

This time, for six out of ten individuals in the sample, the weekly weight loss is less than 2 kg. This obviously makes the alternative hypothesis (7_a) less credible.

However, let us use the hypothesis testing, as before. This time, the sample mean

$$\frac{1}{10}(1.6 + 2.8 + 1.6 + 3.0 + 1.9 + 1.6 + 1.7 + 2.6 + 2.4 + 1.8) = 2.1 \text{ (kg)}$$

is still consistent with the assertion of the famous Dr. C. The sample variance $s^2 \approx 0.2978$ ($s \approx 0.5457$) leads us to the computed value

$$t \approx \frac{2.1 - 2}{0.5457 / \sqrt{10}} \approx 0.5795$$

which is less than $t_{0.05}$. We fail to reject the null hypothesis, even for the "big" significance level $\alpha = 0.05$!

Moreover, the p-value (computed by means of the special function TTEST in *Microsoft Excel*) is obtained as 0.2622. Thus the risk of accepting the alternative hypothesis as true, based on this particular sample, is high enough!

Let us draw some general conclusions about pairs of hypotheses similar to (7_a)-(7_0).

Such an alternative hypothesis (H_a) involves – as a single parameter – the mean μ of a normally distributed population. It is a one-tailed hypothesis

(H_a): $\mu > $ value

and the corresponding null hypothesis takes the form

(H_0): $\mu = $ value .

In the classical approach, once the significance level α is chosen, the critical value $t_\alpha > 0$ that delimits the rejection region $(t_\alpha, +\infty)$ is found from the condition

$$\Theta(t_\alpha) = 1 - \alpha$$

where

$$\Theta(t) = \frac{\Gamma(\frac{N}{2})}{\sqrt{\pi(N-1)} \cdot \Gamma(\frac{N-1}{2})} \int_{-\infty}^{t} \left(1 + \frac{x^2}{N-1}\right)^{-N/2} dx$$

is the distribution function of the Student distribution $t(N-1)$.

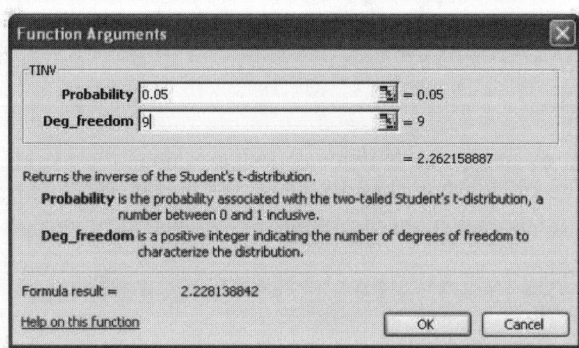

Fig. 4.3. Using TINV in *Microsoft Excel*

Of course, N is the sample size (volume). To reject the null hypothesis, the following value is computed

$$t = \frac{m - \text{value}}{s \big/ \sqrt{N}}$$

where m is the sample mean and s^2 is the sample variance. In case t is greater than t_α, the null hypothesis is rejected.

When using *Microsoft Excel*, the critical value t_α, which depends on the sample size N, is found by means of the function TINV. Its arguments are:

- The significance level α (identified as "Probability", see Figure 4.3 above), and
- The number of degrees of freedom, which is $N-1$.

Example 2. Suppose the alternative hypothesis is (6_a) above:

(6_a)   The average health status of patients, after a standard treatment, is better than before the treatment.

However, it is very difficult to represent, by a single numerical value, the health status of a patient. Let us be more specific, considering only hypotensive patients under the action of an anti-hypotensive drug, and let us evaluate the health status of a patient by his heart rate expressed in beats/minute (b/m).

Hence the pair (6_a)-(6_0) is replaced by

($6'_a$)   The average heart rate of hypotensive patients increases after a drug administration,

res.

($6'_0$)   The average heart rate of hypotensive patients, after a drug administration, does not suffer any change.

Formally, we express the hypotheses above by:

($6'_a$) $\mu_a > \mu_b$

($6'_0$) $\mu_a = \mu_b$

where μ_a, res. μ_b represent the average heart rate after, res. before the drug administration.

The data come naturally in pairs; more precisely, for each patient we measure the hearth rate before (x_b) and after (x_a) the drug administration.

Of course, we could compute the difference $d = x_a - x_b$ and consider the drug to be efficient to our patient when $d > 0$, inefficient when $d = 0$ (i.e. if no change is detected) and harmful when $d < 0$. In fact we are testing the efficiency of our anti-hypotensive drug. Denoting by δ the average difference, the hypothesis testing above is replaced by

(6"$_a$) $\delta > 0$

(6"$_0$) $\delta = 0$

which is exactly the situation treated in Example 1, provided the differences d are distributed normally.

If we suppose that the heart rate of hypotensive patients, either before or after the drug administration, is distributed normally, i.e. is of type $N(\mu_b, \sigma_b^2)$ res. $N(\mu_a, \sigma_a^2)$, then it follows that the differences d are normally distributed, with mean $\mu_a - \mu_b$. The variance of differences is unknown and is estimated by the sample variance s^2. Because we accept *ab initio* that (6'$_0$) is true, the distribution of the differences d is approximately of type $N(0, s^2)$.

Most of the differences are concentrated around 0. Once a significance level α has been chosen, the critical value $t_\alpha > 0$ that delimits the rejection region $(t_\alpha, +\infty)$ is obtained exactly as in Example 1, by means of the Student distribution $t(N-1)$.

Consider the following data (from [Daly, Bourke and McGilvray 1991, p.113]) obtained from a sample of size 8:

Individual	Before (b/m)	After (b/m)	Difference
1	58	66	+8
2	65	69	+4
3	68	75	+7
4	70	68	-2
5	66	73	+7
6	75	75	0
7	62	68	+6
8	72	69	-3

The computed value will be obtained by using the formula

$$t = \frac{m}{s / \sqrt{N}}$$

where m is the sample mean of differences. Here $m = 3.375$, $N = 8$, and $s \approx 4.4058$. Hence $t \approx 2.1667$.

The decision will be taken after comparing this value t to the critical value t_α. However, we fail to reject the null hypothesis even for $\alpha = 0.05$ (because $t_{0.05} \approx 2.3646 > t$).

Example 3. Suppose the alternative hypothesis is (5$_a$) above and, of course, the null hypothesis is (5$_0$):

(5$_a$)   The average salary of male physicians is higher than the average salary of female physicians,

(5_0) The average salary of male physicians coincides with the average salary of female physicians.

Let us rewrite the hypotheses in a more abstract form:

($5'_a$) $\mu_m > \mu_f$

($5'_0$) $\mu_m = \mu_f$

where μ_m res. μ_f represent the average salary of male physicians, res. the average salary of female physicians.

Of course, we start by accepting the null hypothesis as true. We will suppose – *ab initio* – that both populations are distributed normally, i.e. are of type $N(\mu_m, \sigma_m^2)$ res. $N(\mu_f, \sigma_f^2)$.

In Step 2 of a classical hypothesis testing, we have to choose a statistic adapted to the concrete situation. In choosing this statistic we should be aware that in fact two disjoint samples will be selected, one from the population of (salaries of) male physicians, the other from the population of (salaries of) female physicians.

These two samples are, generally, not equal in size. Let us denote by:

– N_m the volume of the sample extracted from the population of male physicians (i.e. from the respective salaries),
– M_m the sample mean of these salaries, and
– S_m^2 their sample variance.

On the other hand, denote by:

– N_f the volume of the sample extracted from the population of (salaries of) female physicians,
– M_f the sample mean of these salaries,
– S_f^2 their sample variance.

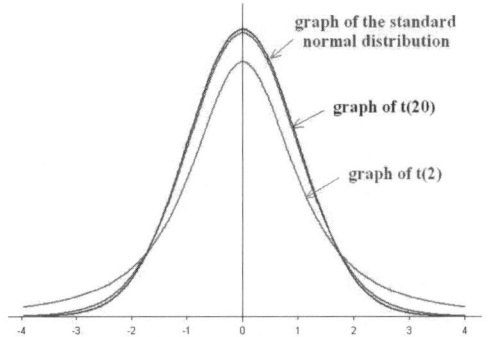

Fig. 4.4. Standard normal compared to Student distributions

A distinction will be made between "small" and "large" samples. Any sample with more than 30 individuals will be considered as "large", otherwise as "small".

The statistic we choose depends strongly on the size of the two samples. When both samples are "large", the formula

$$Z = \frac{M_m - M_f}{S_z}, \text{ where } S_z^2 = \frac{1}{N_m} S_m^2 + \frac{1}{N_f} S_f^2$$

describes Z as a standard normal distribution $N(0, 1)$.

In the other case, i.e. when at least one of the samples is "small", the formula

$$T = \frac{M_m - M_f}{S_t}, \text{ where } S_t^2 = \left(\frac{1}{N_m} + \frac{1}{N_f} \right) \cdot \frac{(N_m - 1)S_m^2 + (N_f - 1)S_f^2}{N_m + N_f - 2}$$

describes T as a Student distribution $t(N_m + N_f - 2)$ – see [Larson 1973].

Remember that for $N \geq 30$ the Student distribution $t(N)$ is approximately the standard normal distribution (see Figure 4.4 above).

Thus, once a significance level α has been chosen, either $N(0, 1)$ or $t(N_m + N_f - 2)$ will be used in order to obtain the critical value – either z_α or t_α – that determines the rejection region.

Notice that in *Microsoft Excel* the function NORMSINV will help us in the first situation. Its argument will be $1 - \alpha$. Of course, in the second situation we use TINV (with α as the argument).

Suppose the two samples are as follows:

Male physician	Salary ($)	Female physician	Salary ($)
1	8105	1	*74410*
2	6719	2	5452
3	7909	3	3814
4	4420	4	4381
5	6214	5	3995
6	9407	6	4944
7	4828	mean m_f	16166
8	6689	variance s_f^2	28540.1
9	7274	size N_f	6
10	8351		
mean m_m	6991.6		
variance s_m^2	1560.2		
size N_m	10		

Attention, if the computed means (i.e. the averages) of the samples are not consistent with the alternative hypothesis, the testing should stop immediately!

Suppose the typing error (7410 instead of 74410) has been detected and corrected. Now $m_m > m_f$, i.e. the computed means are consistent with the alternative hypothesis. The computed value will be obtained by the formula (notice both samples are "small sized"):

$$t = \frac{m_m - m_f}{s_t} \quad \text{where} \quad s_t^2 = \left(\frac{1}{N_m} + \frac{1}{N_f} \right) \cdot \frac{(N_m - 1)s_m^2 + (N_f - 1)s_f^2}{N_m + N_f - 2}$$

and the decision will be taken accordingly.

Of course, there is heavy computing involved that needs some programming work. This is why classical hypothesis testing is far from being widespread.

In general, consider typical alternative hypotheses concerning differences of means. Two kinds of hypotheses are possible:

One-tailed (H_1) : $\mu_1 > \mu_2$,

Two-tailed (H_1) : $\mu_1 \neq \mu_2$.

(The case $\mu_1 < \mu_2$ is exactly the former one-tailed, with a reversed order of the populations.)

In the one-tailed case, once the significance level α has been chosen, the rejection region $(r, +\infty)$ is determined, as in Examples 1-3, from the condition

$$\Phi(r) = 1 - \alpha$$

Φ being an adequate distribution function.

In the two-tailed case, the rejection region is a union $(-\infty, -r) \cup (r, +\infty)$ where $r > 0$ is determined from the condition

$$\Phi(r) = 1 - \frac{\alpha}{2}.$$

The distribution function Φ is either of normal type, or of Student type. In fact, theoretical reasoning identifies several cases.

The variances of the two populations, σ_1^2 res. σ_2^2, are known. In this case the statistic used

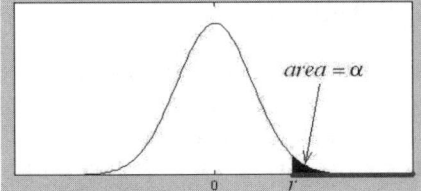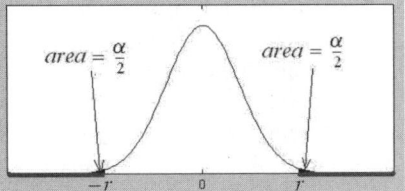

Fig. 4.5. Rejection region for α : one-tailed (left), two-tailed (right)

$$z = \frac{m_1 - m_2}{\sigma} \quad \text{where } \sigma^2 = \frac{1}{N_1}\sigma_1^2 + \frac{1}{N_2}\sigma_2^2$$

follows a standard normal distribution $N(0, 1)$. Of course m_1 and m_2 represent the respective sample means.

The variances of the two populations are unknown, and are replaced by the corresponding sample variances s_1^2 res. s_2^2. However, two exclusive situations should be taken into account.

The homoscedastic case: the unknown variances σ_1^2 and σ_2^2 are equal. In this case the statistic used is

$$t = \frac{m_1 - m_2}{s_p \sqrt{\dfrac{1}{N_1} + \dfrac{1}{N_2}}}$$

where s_p^2 is a pooled estimate of the common population variance, given by

$$s_p^2 = \frac{(N_1 - 1)s_1^2 + (N_2 - 1)s_2^2}{N_1 + N_2 - 2}.$$

In this case the statistic t follows a Student distribution $t(N_1 + N_2 - 2)$.

The heteroscedastic case: the unknown variances σ_1^2 and σ_2^2 are unequal. In this case the statistic used is

$$t = \frac{m_1 - m_2}{s} \quad \text{where } s^2 = \frac{1}{N_1}s_1^2 + \frac{1}{N_2}s_2^2$$

and follows also a Student distribution.

Of course, it is supposed that both populations are normally distributed. However, the results are approximately correct provided the distributions of the populations are not too far from normal.

Let us point out that a Student distribution $t(N)$ with $N \geq 30$ is approximately a standard normal.

4.5 Comparing Means, the Practical Approach

The discussion above shows the difficulties of performing a classical hypothesis test, when comparing means. All the computation is done after the significance level α has been chosen, and after the data from a sample has been extracted.

As exemplified above in Example 1, for a given sample there exists a particular significance level α^*, such that the corresponding critical value (either t_{α^*}, or z_{α^*}) coincides with the computed value. This α^* is the smallest significance level that allows us to accept the alternative hypothesis (by rejecting the null hypothesis), based

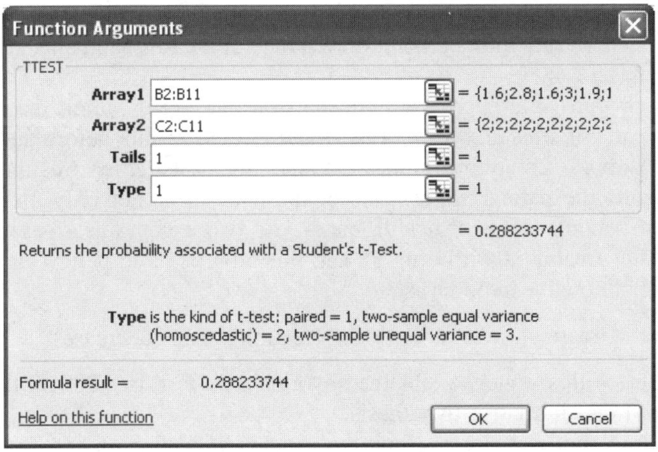

Fig. 4.6. Controlling TTEST in *Excel*

on the given sample. It is called p-value, and it is interpreted as the risk of accepting the alternative hypothesis as true (when in fact the null hypothesis is true).

Modern software eliminates the burden of heavy computation, and allows for the reversing the philosophy underneath hypothesis testing. Instead of choosing at the beginning the significance level (α) and then doing a lot of computation, maybe it is better to first directly compute the p-value of the alternative hypothesis, then to accept it or not as true, depending on how comfortable we feel with the obtained p-value.

This idea is supported in using *Microsoft Excel* by the function TTEST. Its four arguments are, in order (see Figure 4.6):

1) The domain Array1 containing the data extracted from the first sample;
2) The domain Array2 containing the data extracted from the second sample;
3) A numeric (in fact Boolean) parameter Tails, whose value is 1 in case the alternative is one-tailed, res. 2 if two-tailed;
4) A second parameter Type, whose value is 1 in case the samples are paired, 2 if the samples are unpaired but homoscedastic, and 3 if the populations are known as heteroscedastic.

Let us mention here that – in practice – no medical examples exist for which homoscedasticity of populations' is known. For unpaired samples the second parameter should be chosen 3.

The function TTEST gives us directly the p-value of the alternative hypothesis. However, before using TTEST, a preliminary check should be done (otherwise we may draw wrong conclusions). Namely, we should check, by using the function AVERAGE on both domains, if the sample means are correctly ordered.

4.6 Paired and Unpaired Tests

Let us begin this section by considering two formal datasets (see Figure 4.7 below). Notice the last value of each data set is possibly an outlier; it is much larger than the other data from the respective sets.

These outliers raise the respective averages with approx. 25%. However, the order is not influenced. (Even without them, the average of dataset 1 remains larger than the average of dataset 2.)

Consider two possible approaches. In the first one, let us admit the data are obtained from patients treated with a drug D and are lab results before and after treatment (for example, creatinine values). Value decreased after treatment signifies improvement of the patient status. Hence, the data show not only an average improvement of patients status of health, but – with two exceptions – case-by-case improvement after treatment with drug D, and this advises us to believe in the truth of the following alternative hypothesis:

(P_a):   after a treatment with drug D, the creatinine value decreases.

The p-value of this sentence, obtained by a paired t-test, is 0.00010, thus confirming the truth of the alternative hypothesis.

In the second approach, let us admit the data is obtained from two distinct populations, for example the first set is obtained from patients treated with placebo, the

	A	B	C	D	E
1	No.	Dataset 1	Dataset 2		
2	1	148	150		
3	2	150	140	TTEST(B2:B24,C2:C24,1,1)=	0.000100827
4	3	152	150	TTEST(B2:B24,C2:C24,1,3)=	0.408010531
5	4	156	155		
6	5	164	158		
7	6	170	160		
8	7	171	159		
9	8	179	157		
10	9	181	165		
11	10	190	167		
12	11	190	204		
13	12	200	169		
14	13	210	172		
15	14	215	170		
16	15	225	210	TTEST(B2:B23,C2:C23,1,1)=	0.000161761
17	16	238	216	TTEST(B2:B23,C2:C23,1,3)=	0.246756138
18	17	260	250		
19	18	290	279		
20	19	325	312		
21	20	360	356		
22	21	395	325		
23	22	410	397		
24	23	1284	1271		
25	average=	272.30	256.17		
26	average22=	226.32	210.05		

Fig. 4.7. Data sets compared as paired res. unpaired

second set from patients treated with drug D. The smaller average of data set 2 (compared to that of data set 1) indicate an overall efficiency of drug D and advise us to believe in the truth of the following alternative hypothesis:

(U_a):   treatment with drug D is efficient (compared to lack of treatment).

The p-value of this sentence, obtained by an unpaired t-test, is 0.4080. Such a value does not confirm the truth of the alternative hypothesis!

Thus, the same data lead to different conclusions, depending in an essential way of the context the data are obtained.

The same discrepancy appears after renouncing to outliers!

4.7 Example: Comparing Proportions

In Examples 1-3 it was assumed that the populations involved were distributed normally. The pair of hypotheses (1_0)-(1_a) is treated exactly as the pair (5_0)-(5_a) provided the intelligence quotient (IQ) is assimilated to a normal distribution with mean 100 and standard deviation 15.

However, in some situations the populations involved are definitely not normally distributed.

Consider for example the population of patients treated with drug A – see the pair of hypotheses (4_0)-(4_a) above. We can reasonably assume that each patient either recovered (tag 1) or not (tag 0). If we randomly select a sample of N patients treated with drug A, we may assume that the individuals of the sample are (independent) Bernoulli distributions!

Of course, the parameter π_A characterizing such a random variable is interpreted as the probability of "success", i.e. the proportion of a recover. The proportion of recovered people found in the chosen sample gives its natural estimation.

Example 4. Suppose the pair alternative/null hypothesis is as follows:

(4_a) The proportion of recovered people treated with drug A is larger than the proportion of recovered people treated with drug B (i.e. drug A is better than drug B)

(4_0) The proportion of recovered people after treatment with drugs A res. B are the same (i.e. drug A and B are equivalent).

Formally, these hypotheses are rewritten as follows:

($4'_a$) $\pi_A - \pi_B > 0$

($4'_0$) $\pi_A - \pi_B = 0$

where π_A res. π_B represent the respective proportions of recovering.

Let us follow first the classical approach of hypothesis testing. A sample of volume N_A extracted from the population of people treated with drug A is in fact a sequence $X_1, X_2, ..., X_{N_A}$ of (independent) Bernoulli distributions of type $Be(\pi_A)$. The number of recovered people is the sum $X_1 + X_2 + ... + X_{N_A}$ and, as such,

is a binomial distribution $b(N_A, \pi_A)$. The sample mean
$P_A = \dfrac{1}{N_A}(X_1 + X_2 + ... + X_{N_A}) = \dfrac{1}{N_A} b(N_A, \pi_A)$ is a statistic that expresses
the proportion of recovered people from those treated with drug A.

Analogously, a sample of volume N_B extracted from the population of people
treated with drug B is in fact a sequence $Y_1, Y_2, ..., Y_{N_B}$ of Bernoulli distributions of
type $Be(\pi_B)$ and the number of recovered people $Y_1 + Y_2 + ... + Y_{N_B}$ is a binomial
distribution $b(N_B, \pi_B)$. Of course, the sample mean
$P_B = \dfrac{1}{N_B}(Y_1 + Y_2 + ... + Y_{N_B}) = \dfrac{1}{N_B} b(N_B, \pi_B)$ is a statistic that expresses the
proportion of recovered people from those treated with drug B.

It is well known that, in general, the binomial distribution $b(n, p)$ is approximately normal with mean $\mu = np$ and variance $\sigma^2 = np(1 - p)$, provided
$0.1 < p < 0.9$, $np \geq 5$ and $n(1 - p) \geq 5$.

Suppose the conditions $0.1 < \pi_A < 0.9$, $0.1 < \pi_B < 0.9$, $N_A \pi_A \geq 5$,
$N_A(1 - \pi_A) \geq 5$, $N_B \pi_B \geq 5$ and $N_B(1 - \pi_B) \geq 5$ are fulfilled. Then the random
variable $\dfrac{1}{N_A} b(N_A, \pi_A)$ is approximately (normal) of type

$N\left(\pi_A, \dfrac{\pi_A(1 - \pi_A)}{N_A}\right)$, and $\dfrac{1}{N_B} b(N_B, \pi_B)$ is approximately of type

$N\left(\pi_B, \dfrac{\pi_B(1 - \pi_B)}{N_B}\right)$. Hence the difference

$$D = P_A - P_B$$

which expresses the difference of proportions of recovered people, will be approximately of type

$$N\left(\pi_A - \pi_B, \dfrac{\pi_A(1 - \pi_A)}{N_A} + \dfrac{\pi_B(1 - \pi_B)}{N_B}\right).$$

Let us explore the consequences of a true null hypothesis (4"$_0$) $\pi_A = \pi_B = \pi$. It
follows that $D = P_A - P_B$ is approximately of type $N\left(0, \pi(1 - \pi)\left(\dfrac{1}{N_A} + \dfrac{1}{N_B}\right)\right)$.

Thus, in order to find a suitable statistic, we need a pooled estimate of π, the percentage of recovered people in those treated with drugs (either A or B). This is obtained as the statistic

$$P = \frac{1}{N_A + N_B}(X_1 + X_2 + ... + X_{N_A} + Y_1 + Y_2 + ... + Y_{N_B}).$$

Our statistic to be used in the hypothesis testing will be

$$Z = \frac{P_A - P_B}{\sqrt{P(1-P)\left(\dfrac{1}{N_A} + \dfrac{1}{N_B}\right)}}$$

the notation Z indicating that it is a standard normal distribution, i.e. of type $N(0,1)$.

Suppose the data obtained from samples are as follows:

	Sample A	Sample B	Both samples
Total patients	$N_A = 80$	$N_B = 75$	$N_A + N_B = 155$
Recovered	55	40	95
Percentage of recovered	$P_A = 68.75\%$	$P_B = 53.33\%$	$P \approx 61.29\%$

From here, the z-score is $z \approx 1.9692$, and this corresponds to a p-value $\alpha^* \approx 0.02446$ (see Figure IV.8). This is also known as the Mid-p value (for example it is identified as such in *Epi Info 2004*) and is interpreted according to our risk adversity.

Obviously, before computing the z-score, we have to check whether or not the two percentages P_A and P_B are in the correct relation; if not, the testing will stop.

	A	B	C	D	E	F
1		Drug A	Drug B	Row totals		
2	Recovered	55	40	95		
3	Not recovered	25	35	60		
4	Column totals	80	75	155		
5	Rec. proportions	0.6875	0.53333333	0.612903226		
6		=B2/B4	=C2/C4	=D2/D4		
7						
8	z-score	1.969221	=(B6-C6)/SQRT(D6*(1-D6)*(1/B4+1/C4))			
9	p-value	0.024464	=1-NORMSDIST(B8)			

Fig. 4.8. Computing the p-value in *Excel*

When using *Microsoft Excel*, the p-value α^* is given by a formula

$$= 1 - \text{NORMSDIST}(x)$$

where x denotes the coordinates of the cell where the z-score has been computed.

It is customary (and well suited when using *Microsoft Excel*) to present the outcomes from samples in a contingency table, such as:

Treated with / Number of	Drug A	Drug B
Recovered patients	55	40
Not recovered patients	25	35

Do not forget the conditions $N_A \pi_A \geq 5$ etc. However, π_A and π_B are unknown and will be estimated by P_A res. P_B. The values $N_A P_A$ etc. are exactly the numeric values in the cells of the contingency table above. Thus the classical approach of hypothesis testing may be used only when all numeric components of the contingency table have values at least 5.

There is another method to treat such data, known as the chi-square test. This method compares two 2-values random variables, and evaluates their statistical independence.

The statistical independence of two random variables V, W means that

$$P(V = v \wedge W = w) = P(V = v) \cdot P(W = w)$$

for each values v of V and w of W.

In our case, the variable V is the "Drug" and its values are $v \in \{"Drug\ A", "Drug\ B"\}$; on the other hand, W is the "Recovery status of patients", with values $w \in \{"Recovered", "Not\ recovered"\}$.

If the probabilities are estimated by the relative frequencies (using data from samples), then the statistical independence of V and W corresponds to the linear dependence of rows (or of columns) of the extended contingency table:

	$\ldots$	w	$\ldots$	Row totals
$\ldots$	$\ldots$	$\ldots$	$\ldots$	$\ldots$
v	$\ldots$	N_{vw}	$\ldots$	$N_{v\bullet}$
$\ldots$	$\ldots$	$\ldots$	$\ldots$	$\ldots$
Column totals	$\ldots$	$N_{\bullet w}$	$\ldots$	$N_{\bullet\bullet}$

where:

N_{vw} is the number of cases for which $V = v$ and $W = w$,

$N_{v\bullet}$ is the number of cases for which $V = v$, i.e. $N_{v\bullet} = \sum_{w \in W} N_{vw}$,

$N_{\bullet w}$ is the number of cases for which $W = w$, i.e. $N_{\bullet w} = \sum_{v \in V} N_{vw}$,

$N_{\bullet\bullet}$ is the total number of cases, i.e. $N_{\bullet\bullet} = \sum_{v \in V} \sum_{w \in W} N_{vw}$.

This linear dependence means that

$$N_{vw} = \frac{N_{v\bullet} N_{\bullet w}}{N_{\bullet\bullet}} \text{ for each values } v \text{ of } V \text{ and } w \text{ of } W$$

or that the value of the expression

$$X^2 = \sum_{v \in V} \sum_{w \in W} \frac{N_{\bullet\bullet}}{N_{v\bullet} N_{\bullet w}} \left(N_{vw} - \frac{N_{v\bullet} N_{\bullet w}}{N_{\bullet\bullet}} \right)^2$$

is 0.

In our case, the extended contingency table is:

	Drug A	Drug B	Row totals
Recovered	55	40	95
Not recovered	25	35	60
Column totals	80	75	155

and the linear dependence of rows (or of columns), i.e. the statistical independence of variables "Drug" and "Recovery status of patients" means exactly that the null hypothesis $(4'_0)$ $\pi_A = \pi_B$ is true.

The formula above, which gives the so-called the "X square statistic", measures in some way how untrue the null hypothesis is. Big values of X^2 will determine us to reject the null hypothesis.

It is well known in Statistics that X^2 has approximately a distribution of type $\chi^2 \big((r-1)(c-1) \big)$, where r is the number of distinct values of V and c is the number of distinct values of W. In our case $r = c = 2$, hence X^2 is of type $\chi^2(1)$.

Now, the graph of the chi-square distribution could be used to reject/not reject the null hypothesis.

All the above considerations are drastically simplified when using *Microsoft Excel*! Indeed, we find here a built-in function called CHITEST, having two arguments:

- the rectangular domain (Actual_range) containing the contingency table,
- the rectangular domain (Expected_range) containing theoretical data that correspond to the null hypothesis, i.e. computed using the formula

$$N_{vw} = \frac{N_{v\bullet} N_{\bullet w}}{N_{\bullet\bullet}}.$$

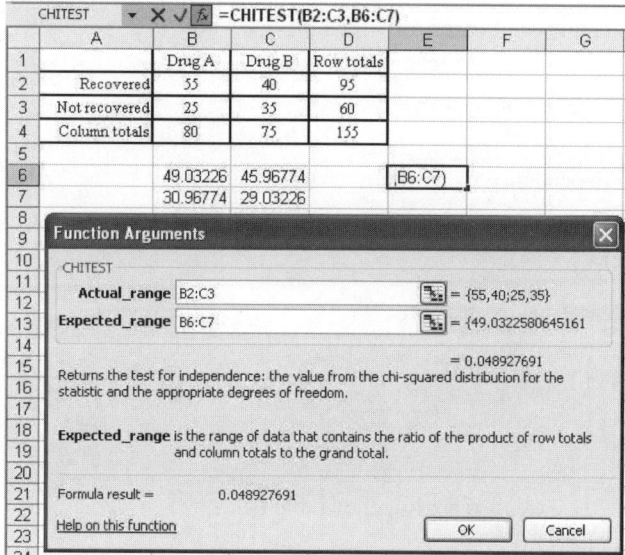

Fig. 4.9. Computing p-value in *Excel*

This function returns the p-value α^*, which may be interpreted by anyone according to his/hers risk adversity.

In our case we obtain (see Figure 4.9) $\alpha^* = 0.0489...$

It is worth noting that the two methods we used above give different p-values (0.02446 res. 0.0489). This is not surprising. There are several reasons: the use of intermediate estimations of proportions, the approximation of "true" distributions – of the used statistics – by others, of normal or chi-square type etc. Even the alternative hypothesis is different.

The pair of hypotheses (2_a)-(2_0) is similar to (4_a)-(4_0). The two methods presented above are well suited to ascertain opinions or findings about incidence of diseases, similar to (2_a).

However, the chi-square distribution is used also to ascertain opinions such as (2_0) or (4_0), i.e. opinions expressing equality or coincidence. The respective tests are known as goodness-of-fit tests.

4.8 Goodness-of-Fit: Chi-Square

Census data in many countries show that the proportion of girls at birth is slightly greater than 0.5, usually 0.51. It is unanimously accepted that the (genders of) newborn children may be considered as independent random selections from a sequence of Bernoulli distributions with parameter 0.51. This implies that the gender of the second newborn in each family is statistically independent of the gender of the first.

Is this general opinion sustained by statistical data? If the independency hypothesis were true, then the number of girls in 4-children families were a binomial distribution $b(4; 0.51)$ described (approximately) as follows:

Number of girls	0	1	2	3	4
Probability	0.058	0.240	0.374	0.260	0.068

If we collect data from – let us say 1000 four-children families – then we expect the following frequencies to be found:

Number of girls	0	1	2	3	4
Frequency of families	58	240	374	260	68

What if we will observe a different number of frequencies? Of course, large differences will force us to revise our opinions.

The situation above is an example of a so-called multinomial experiment. In general, such an experiment is characterized by:

a) A number of N independent observations that are carried out, each observation falling into one of K categories (denoted below by $C_1, C_2,..., C_K$). The observed category frequencies are denoted by $O_1, O_2,..., O_K$. Obviously, $O_1 + O_2 + ... + O_K = N$ (i.e. the sample size);

b) K probabilities. The probability p_k that an (arbitrary) observation falls into category k is known and does not change from one observation to another ($k = 1, 2,..., K$). Of course, $p_1 + p_2 + ... + p_K = 1$. The expected frequencies are computed using the formula

$$E_k = N \cdot p_k, (k = 1, 2,..., K). \text{ Of course, } E_1 + E_2 + ... + E_K = N.$$

Does the observed data confirm the expectations? Of course, large discrepancies between observed data $\{O_k\}$ and expected data $\{E_k\}$ will stand against the independency hypothesis. However, the main problem is: how discrepancy should be evaluated?

The solution proposed by Karl Pearson[2] in 1900 uses the number

$$X^2 = \sum_k \frac{1}{E_k}(O_k - E_k)^2$$

as a distance. From a statistical point of view, X^2 is a statistic – i.e. a formula involving data extracted from a sample – whose distribution is approximately that of $\chi^2(K-1)$ and a chi-square test could be performed. The threshold between "small discrepancies" and "large discrepancies" may be interpreted in terms of the $\chi^2(K-1)$ distribution.

---

[2] Karl Pearson (1857-1936), English mathematician, founder of the statistical journal *Biometrika*.

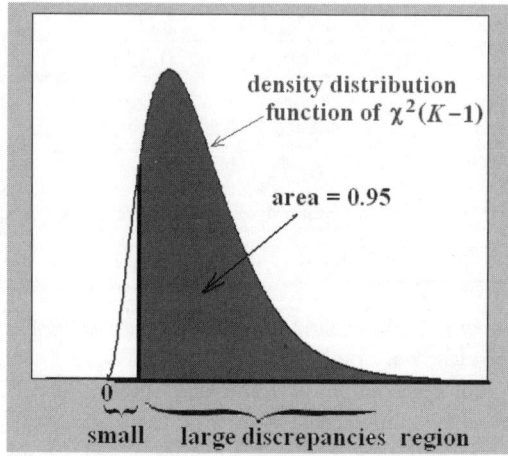

Fig. 4.10. Using chi-square distribution

In the Figure 4.10 this threshold was fixed according to a level of significance 0.95.

In practice, by using *Microsoft Excel*, we could make use of the function CHITEST and to interpret its return (i.e. the "p-value") in the reversed way. Values (very) near to 1 are interpreted as follows: "the observed data confirm the hypothesis".

For example, suppose that the data collected from the 1000 four-children families are as follows:

Number of girls k	0	1	2	3	4	Total
Number of four-children families having k girls (observed frequency O_k)	56	233	377	268	66	1000
Expected frequency E_k	58	240	374	260	68	
Difference $(O_k - E_k)$	–2	–7	3	8	–2	

The obtained p-value (see Figure IV.11 below) is 0.963, which means that data obtained from the chosen sample supports the opinion that the gender of the second newborn in a family is statistically independent of the gender of the first newborn in that family.

The first application of the chi-square goodness-of-fit test dates back to 1901, when a very important theory in genetics was confirmed (because it was only in 1900 when an important paper of Gregor Mendel – published as early as 1865! – came into light). Mendel noticed that some characteristics of garden peas plants may disappear in all direct offspring, but may reappear in some second-generation offspring.

A plausible explanation is based on the idea that a "genetic" characteristic of individuals is determined in each individual by a pair of gametes, which are inherited one from the father, the other from the mother. Now, if each of the two genes may take

	A	B	C	D	E	F	G
1	Number of girls k	0	1	2	3	4	Total
2	Number of four-children families having k girls (observed frequency O_k)	66	268	377	233	56	1000
3	Expected frequency E_k	68	260	374	240	58	
4							
5	0.96282183	=CHITEST(B2:F2,B3:F3)					

Fig. 4.11. Confirming a hypothesis

only two values (called alleles) A and a, then a given individual is either homozygous (i.e. the two gametes of the gene are identical: AA or aa) or heterozygous (i.e. the two gametes are different). When two homozygous individuals of different alleles are crossed, all the direct offspring are identical heterozygous (this is the "law of uniformity").

Suppose the determined "genetic" characteristic is the stature, with two possible values: tall res. short. The hypothesis is that individuals with gametes AA or Aa (= aA) are tall, and individuals with gametes aa are short (i.e. the allele A is "dominant").

In a population where alleles are equally distributed, the proportion of homozygous individuals is $\frac{2}{4} = 0.5$, and the proportion of tall individuals is $\frac{3}{4} = 0.75$.

If the proportion of the dominant allele is π, then the proportion of homozygous individuals is $\pi^2 + (1-\pi)^2$, and the proportion of tall individuals is $\pi^2 + 2\pi(1-\pi)$. Moreover, the proportions of the three genotypes AA, Aa (= aA) and aa are, respectively:

$$\pi^2,\ 2\pi(1-\pi),\ (1-\pi)^2.$$

The Hardy[3] – Weinberg principle states that the proportions of the different genotypes remain constant from one generation to the next one (i.e. the population is in equilibrium).

To test whether the population under study is in equilibrium, suppose amongst 1000 individuals, selected at random, the following genotype frequencies are observed:

Genotype	AA	Aa (= aA)	aa	Total
Observed frequency	799	188	13	1000

The incidence of the A allele (which is an estimator of π) is obviously

$$p = \frac{2 \cdot 800 + 1 \cdot 185}{2 \cdot 1000} = 0.8925.$$

---

[3] Godfrey Harold Hardy (1877-1947), famous English mathematician.

Hence the following expected "frequencies" are computed:

Genotype	AA	Aa (= aA)	aa
Expected frequency	796.56 $=1000 \cdot p^2$	191.89 $=1000 \cdot 2p(1-p)$	11.56 $=1000 \cdot (1-p)^2$

The chi-square test gives a p-value of 0.5251. This value does not support the hypothesis that the population is in equilibrium. (This should trigger an investigation! It is usually assumed that most biological populations are in equilibrium for most genetic characteristics.)

Mendel obtained 556 peas [Cramér 1955] that were classified in four groups, according to two characteristics:

- The shape (values round/angular),
- The color (values yellow/green).

Group	round and yellow	round and green	angular and yellow	angular and green	Total
Observed frequency	315	108	101	32	556

From here he deduced that round and yellow are the values determined by the dominant alleles A res. B. Mendel hypothesized that these alleles (and the corresponding recessive ones a res. b) are equally distributed in the population. Hence we should expect the following:

Group	round and yellow	round and green	angular and yellow	angular and green
Expected proportion	$\dfrac{9}{16}$	$\dfrac{3}{16}$	$\dfrac{3}{16}$	$\dfrac{1}{16}$
Expected frequency	312.75	104.25	104.25	34.75

The chi-square test gives a p-value of 0.9254, which is large enough to confirm Mendel's hypothesis.

A chi-square goodness-of-fit test may be used to confirm a supposed distribution of a population, based on data obtained from a random sample. For example, since Quételet[4] it is widely accepted that most numerical characteristics (such as the height or the weight) of big biological populations are approximately normally distributed. A chi-square test may be used to ascertain the normality.

Usually the numerical data $x_1, x_2, ..., x_N$ ($\in \mathbf{R}$) obtained from a sample of volume N are grouped into K groups (or bins) determined by $K - 1$ separation values $s_1 < s_2 < ... < s_{K-1}$ and observed frequencies O_k are easily computed.

[4] Adolphe Quételet (1796-1874), Belgian statistician. The body mass index is known also as the Quételet index.

(More precisely, the value x_i is placed into bin k if

$$s_{k-1} < x_i \le s_k, \text{ where } s_0 = -\infty \text{ and } s_K = +\infty .)$$

Suppose the population is normally distributed, with mean μ and variance σ^2. It is well known that μ is estimated by the average $m = \dfrac{x_1 + x_2 + ... + x_N}{N}$ and σ is estimated by the standard deviation $s = \sqrt{\dfrac{(x_1 - m)^2 + (x_2 - m)^2 + ... + (x_N - m)^2}{N-1}}$.

Once m and s are found, it is easy to obtain an estimate of the probability that a value x falls into the interval $(s_{k-1}, s_k]$:

$$p_k = \frac{1}{s\sqrt{2\pi}} \int_{s_{k-1}}^{s_k} \exp\left(-\frac{(x-m)^2}{2s^2} \right) dx$$

and from here the estimated frequency $E_k = p_k \cdot N$ of the bin k.

In *Microsoft Excel* the function FREQUENCY is used to obtain the observed frequencies O_k, then the functions AVERAGE and STDEV are used to obtain the estimations m res. s. As for the estimated probabilities p_k, NORMDIST is available. Finally, CHITEST will return the p-value.

	A	B	C	D	E	F	G	H	I
1	values	m	0.49089			s	0.28965		
2	0.32009		=AVERAGE(A2:A101)			=STDEV(A2:A101)			
3	0.81807								
4	0.44454		separators		observed frequencies				
5	0.99338		m - 3s	-0.37806	0	=FREQUENCY(A2:A101:D5:D11)			
6	0.19679		m - 2s	-0.08841	0				
7	0.90682		m - s	0.20124	20				
8	0.04855		m	0.49089	34				
9	0.0369		m + s	0.78053	27				
10	0.58955		m + 2s	1.07018	19				
11	0.14833		m + 3s	1.35983	0				
12	0.95391				0				
13	0.53427								
14	0.30168				expected frequencies				
15	0.96932			0.13500	=100*NORMDIST(D5,C1,F1,TRUE)				
16	0.89201			2.14001					
17	0.6276			13.59052					
18	0.86144			34.13447					
19	0.44701			34.13447					
20	0.30205			13.59052					
21	0.60405			2.14001					
22	0.54857			0.13500					
23	0.70592		p-value						
24	0.00795		0.07271	=CHITEST(E5:E12,E15:E22)					
25	0.28093								

Fig. 4.12. Checking the random number generator

In Figure 4.12 an Excel spreadsheet is presented. 100 numerical values were randomly generated using the random number generator RAND, and then grouped into 8 bins. The obtained p-value is 0.07271, which does not confirm the normality! This is not surprising, because RAND generates uniformly (not normally) distributed numbers.

Other software produces diagrams that allow visual comparison between a histogram of the sampled data and a graph of the (density of the) estimated normal distribution. The decision – i.e. to accept or to reject the hypothesis that the population is normally distributed – is left to the user.

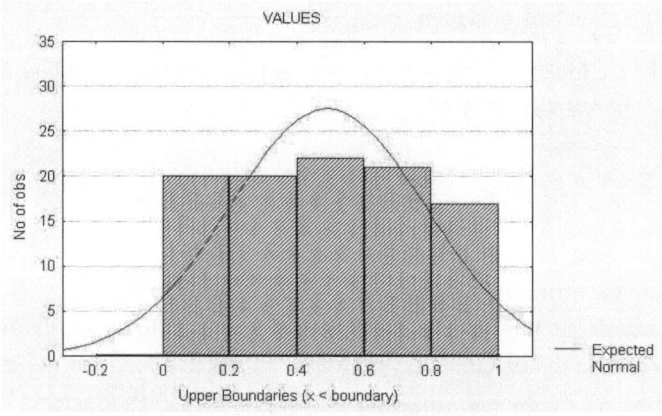

Fig. 4.13. Visual check of normality with *Statistica*

For example, *Statistica* – commercial statistical software developed by StatSoft Inc. – produces, from the generated data above, the diagram in Figure 4.13.

However, a serious error was made in the example above. To understand why, let us remember that each chi-square test is based on the statistic X^2, which has, approximately, a chi-square distribution. The approximation is good only if the expected frequency of each bin is at least 5!

Let us summarize the chi-square goodness-of-fit test. It can be applied for any univariate distribution – either discrete or continuous – for which the distribution function Θ can be computed. It is supposed that:

(H_0): the data follow the specified distribution
      and the alternative
(H_a): the data do not follow the specified distribution.

The real numbers are divided into K bins, each bin being an interval $(s_{k-1}, s_k]$ that contains at least one observation. The statistic used is

$$X^2 = \sum_k \frac{1}{E_k}(O_k - E_k)^2$$

where O_k is the observed frequency for bin k and E_k is the expected frequency for that bin, which is computed by using the formula

$$E_k = N \cdot (\Theta(s_k) - \Theta(s_{k-1})).$$

The statistic X^2 is approximately $\chi^2(K - C)$ distributed, where C is $1 +$ the number of parameters that are to be estimated for the distribution in question.

For a good approximation it is required an expected frequency of at least 5 for each bin. (Any bin with expected frequency less than 5 should be attached to a neighboring bin.)

4.9 Other Goodness-of-Fit Tests

In many practical situations, to be able to use a specific method, some *a priori* conditions regarding the available data should be met. Foe example, a common condition is that a sample comes from a normally distributed population.

Goodness-of-fit tests are tools suitable to confirm that available data follow a specified distribution. What we want to confirm is:

(H_0):   the data follow a distribution fully specified by the distribution function
$F : \mathbf{R} \to [0,1]$

by rejecting

(H_a):   the data do not follow the distribution in question.

Suppose the data obtained from the sample

$$x_1, x_2, ..., x_N$$

has been ordered:

$$x_{(1)} \leq x_{(2)} \leq ... \leq x_{(N)}.$$

In case F is continuous, to apply the Kolmogorov-Smirnov test the following statistic is computed

$$D = \max_{1 < n \leq N} \left\{ F(x_{(n)}) - \frac{n-1}{N}, \frac{n}{N} - F(x_{(n)}) \right\}.$$

In case F is normal, and the volume of the sample is between 10 and 40, to apply the Anderson-Darling test the following statistic is computed:

$$A = \sqrt{-N - \sum_{n=1}^{N} \frac{2n-1}{N} \ln\left\{ F(x_{(n)}) \cdot (1 - F(x_{(N+1-n)})) \right\}}.$$

Accepting a theory, based on evidence collected from a sample, after a goodness-of-fit test, is always a personal decision. (Of course, there is an associated risk!)

To support a subjective decision – when detail and/or time are lacking, the so-called quantile-quantile plot (or q-q-plot) is used. This is a diagram in which the collected data, increasingly ordered

$$x_{(1)} \leq x_{(2)} \leq \ldots \leq x_{(N)}$$

are compared to the data

$$y_1 \leq y_2 \leq \ldots \leq y_N$$

that correspond to the theoretical distribution specified in the null hypothesis (H_0). More precisely,

$$F(y_n) = \frac{n}{N+1} \quad \text{for } n \in \{1, 2, \ldots, N\}.$$

A point in the q-q-plot – see Figure IV.14 for an example – represents a pair $(x_{(n)}, y_n)$. If the original (unsorted) data $\{x_n\}$ are "extracted" from the distribution in question, then all the points lay exactly on the diagonal. Hence, we will accept – subjectively – the null hypothesis as "true" only if all the points are "near" to the diagonal.

4.10   Nonparametric Tests. Wilcoxon/Mann-Whitney

In the previous paragraphs the problem of comparing two populations by taking into account their means or proportions, maybe also their variances, was approached. In other words, we considered the parameters that characterize the populations: the means (μ), the proportions (π) and the variances (σ^2).

Most of the comparison methods used for continuous random variables are based on the "fundamental" hypothesis that some random variables are normally distributed (or, at least, approximately normal distributed). Because of this, they are known in the statistical literature as parametric tests.

There are situations in which either the distribution of the variables is not known, or the normality hypothesis is clearly not respected. In such situations, to compare the population we can use tests that do not suppose anything about the distribution type, i.e. nonparametric tests.

(Obviously, such tests could we used also in case of variables that are normally distributed. However, the obtained results will be less "significant" that those obtained using analogous parametric tests.)

In the most known nonparametric tests, the numeric values of the variables – obtained from the sample – are replaced by their ranks. For this reason these are called rank tests.

Let us present, in what follows, one of the simplest rank test, namely the Wilcoxon test. The initial alternative hypothesis, in a general expression, is as follows:

(H_a):   the distribution of the values of the numeric random variable (of interest) is asymmetrical around 0.

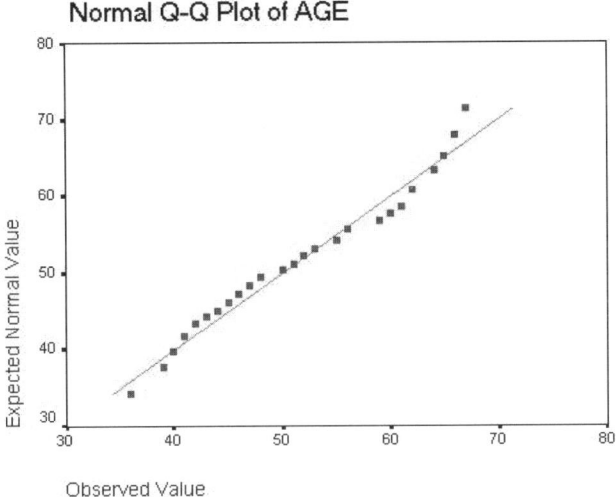

Fig. 4.14. A q-q-plot obtained with *SPSS* (commercial computer software developed by SPSS Inc. (*www.spss.com*)

The opposed null hypothesis is the following:

(H$_0$): the distribution of the values of the numeric random variable (of interest) is symmetrical around 0.

According to the general theory of testing, we will try to "deduce" logical consequences of the truth of null hypothesis, then to see whether the data obtained from sample are "compatible" or not with these consequences.

Let us begin by analyzing numerical data $x_1, x_2, ..., x_n$ obtained from a sample of volume n. Obviously, some of the data are positive, some other are negative, and it is perfectly possible to find some data which are 0. Suppose m data ($m \le n$) are non-zero.

Consider the absolute values $|x_i|$ in increasing order

$$|x_{(1)}| \le |x_{(2)}| \le ... \le |x_{(m)}|$$

then replace each data with its corresponding rank. Denote by T_+ the sum of ranks of positive values and by T_- the sum of ranks of negative values. If hypothesis H$_0$ is accepted as true, then T_+ and T_- are not too different. On the other hand, their sum $T_+ + T_-$ should be equal to the sum of all ranks, i.e. to $\dfrac{m(m+1)}{2}$. We should expect both T_+ and T_- be "nearly" $\dfrac{m(m+1)}{4}$. As T_+ departs from $\dfrac{m(m+1)}{4}$, the null hypothesis becomes implausible and, as a consequence, we are tempted to believe in the truth of the alternative (H$_a$).

The computation of the p-value of the alternative hypothesis is based on the fact that the statistic

$$\frac{T_+ - m(m+1)/4}{\sqrt{m(m+1)(2m+1)/24}}$$

is distributed (at least for "large" values of n) approximately standard normal.

As an example, consider the data in the *Excel* spreadsheet in Figure 4.15 below. Notice that five out of the nine values are positive, the other four are negative (none is zero). Apart of the sign, their order is as follows:

$$0.4 < 0.5 < 0.6 < 1.9 = 1.9 < 2.1 < 3.5...$$

Two of the positive values are equal; hence their ranks will be both $\dfrac{4+5}{2} = 4.5$.

	A	B	C
1	values	sign	rank
2	1.9	+	4.5
3	1.9	+	4.5
4	3.5	+	7
5	0.5	+	2
6	-3.6	--	8
7	-5.2	--	9
8	0.6	+	3
9	-2.1	--	6
10	-0.4	--	1

Fig. 4.15. Sign and rank of values

The effect of the command

MEANS values sign

in *Epi Info* is presented in Figure 4.16. The p-value of the alternative hypothesis, obtained by Wilcoxon test, is 0.0139, sufficiently small to convince us to accept it as true.

Therefore, we could state that the set of five positive values differs "significantly" from the set of four negative values. (Let us notice that by using the classical Student test the computed p-value is 0.0042, three times less. However, how could be sure that all preliminary conditions of normality, needed in order to apply the Student test, are satisfied?)

Remember the t test is used, in general, in the following context:

– There are two samples, extracted respectively from two populations,
– The obtained values from the individuals from samples are numerical,
– We want to confirm that the center of (values of) the first population differs from the center of (values of) the second, and
– We locate the center of a population in its mean.

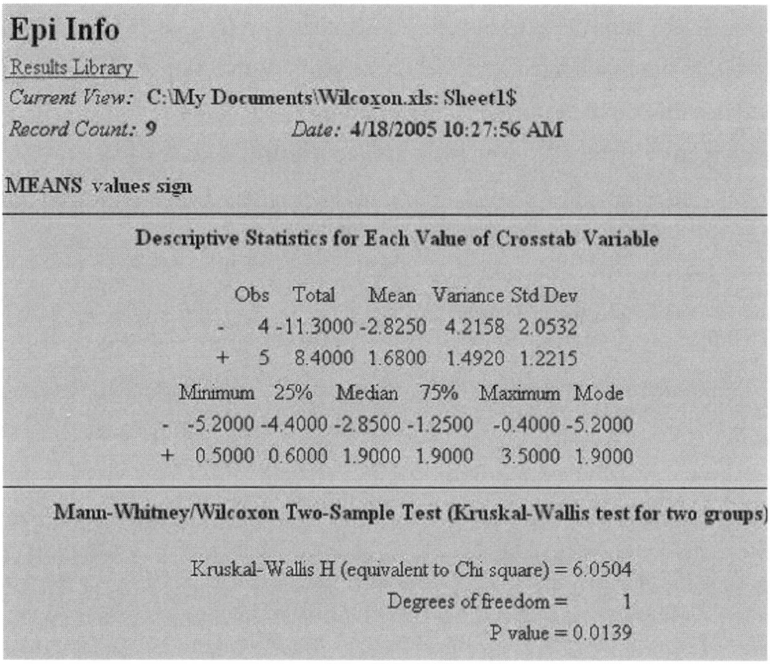

Epi Info

Results Library

Current View: C:\My Documents\Wilcoxon.xls: Sheet1$

Record Count: **9**   *Date:* 4/18/2005 10:27:56 AM

MEANS values sign

Descriptive Statistics for Each Value of Crosstab Variable

	Obs	Total	Mean	Variance	Std Dev
-	4	-11.3000	-2.8250	4.2158	2.0532
+	5	8.4000	1.6800	1.4920	1.2215

	Minimum	25%	Median	75%	Maximum	Mode
-	-5.2000	-4.4000	-2.8500	-1.2500	-0.4000	-5.2000
+	0.5000	0.6000	1.9000	1.9000	3.5000	1.9000

Mann-Whitney/Wilcoxon Two-Sample Test (Kruskal-Wallis test for two groups)

Kruskal-Wallis H (equivalent to Chi square) = 6.0504

Degrees of freedom =   1

P value = 0.0139

Fig. 4.16. Results obtained using *Epi Info*

The center of a population can be localized also in its median, especially in situations when we are interested mostly in the ranks, not in the concrete numerical values.

Numerical values obtained from samples (extracted from populations) appear not only as a result of measurings; transformations, rather arbitrary, of ordinal data could end in numbers. Some examples:

hipo = +1, medium = +2, hyper = +3;

$- - - = -3$, $- - = -2$, $- = -1$, $+ = 1$, $+ + = 2$.

In such situations applying the t test is not justified. However, nothing prevents us to apply non-parametric tests.

Suppose the data extracted from the first population lead to the set of numerical values

$$x_1, x_2, ..., x_{n_1}$$

and, analogously, the data extracted from the second population lead to

$$y_1, y_2, ..., y_{n_2}.$$

According to Wilcoxon idea, let us order the values from the union of two sets, and then attach to each value its rank. (Obviously, the rank is computed in case several values are equal.)

Denote T_1 the sum of ranks obtained from all n_1 values x_i that form the sample extracted from the first population. Analogously, T_2 is the sum of ranks of n_2 values y_j that form the sample extracted from the second population.

The alternative hypothesis, which we want to confirm, is as follows:

(H_a):   the distribution of the values x in the first population differs from the distribution of values y in the second population

and the confirmation will take place by rejecting the null hypothesis:

(H_0):   the distribution of the values x in the first population coincides with the distribution of values y in the second population.

The minimum of the sum of ranks T_1 is $n_1(n_1+1)/2$ and the maximum is $n_1 n_2 + n_1(n_1+1)/2$. On the other hand, if we accept the null hypothesis as true, we expect the sum T_1 be equal to $n_1(n_1+n_2+1)/2$. As T_1 "departs" from this value (towards the extremes $n_1(n_1+1)/2$, res. $n_1 n_2 + n_1(n_1+1)/2$), the null hypothesis becomes less and less plausible. Thus, Wilcoxon test is based on an obvious computing of sums of ranks.

In medical literature another test, the Mann-Whitney test, is often encountered. This test is meant to solve the same problems as the Wilcoxon test. In fact, the two are equivalent.

In short, in the Mann-Whitney test, instead of a sum of ranks computation, all pairs (x_i, y_j) are compared; U_{XY} represents the number of pairs (x_i, y_j) such that $x_i < y_j$ plus one half of the number of pairs such that $x_i = y_j$.

The number U_{XY} is between 0 and $n_1 n_2$. In case the null hypothesis is true, we expect this number be equal $\dfrac{n_1 n_2}{2}$. As U_{XY} "departs" from $\dfrac{n_1 n_2}{2}$, the null hypothesis becomes less plausible.

The relation between Wilcoxon test and Mann-Whitney test is given by the formula

$$U_{XY} = n_1 n_2 + n_1(n_1+1)/2 - T_1$$

which expresses the number U_{XY} (Mann-Whitney) in function of the ranks sum T_1 (Wilcoxon). It is not surprising that in *Epi Info* reports the results are presented together (see Figure 4.16 above).

The Kruskal-Wallis test is nothing else than a generalization of the Wilcoxon test, for the case of more than two samples.

4.11   Analysis of Variance

In the second part of the previous section we analyzed comparatively two groups of the same population, namely the group of individuals treated with drug D respectively the group of individuals treated by placebo. The groups were considered as samples extracted from different populations.

If drug D is prescribed in several different doses, several groups appear.

Sometimes we have to compare more than two populations, or more than two strata of the same populations, and the comparison is made at the level of means. In such situations a generalization of the t test for two populations is used; this generalization known as analysis of variance, or ANOVA test.

From historical point of view, in the first application of the analysis of variance the crops obtained after treatment of soil with different type of fertilizers were compared. Some of the notations and notion used then (such as "mean of treatment") are traditionally maintained.

To explain how analysis of variance is accomplished, let us consider several populations, each population having a mean and a variance (obviously, unknown). From each population a sample is extracted, as follows:

Population 1		Population k		Population K
mean μ_1	$\cdots$	mean μ_k	$\cdots$	mean μ_K
variance σ_1^2		variance σ_k^2		variance σ_K^2
Sample of		Sample of		Sample of
volume n_1		volume n_k		volume n_K
sample mean m_1		sample mean m_k		sample mean m_K
sample variance s_1^2		sample variance s_k^2		sample variance s_K^2

Analysis of variance is conducted for the following null hypothesis

(H_0): no differences between the populations means exist to be rejected, in order to confirm the alternative hypothesis

(H_a): at least two of the means μ_k are different (i.e. at least two of the populations differ in mean).

As usual in hypothesis testing, suppose the null hypothesis is true and deduce logical consequences of this fact. If no differences between the populations means μ_k exist, we should expect the sample means m_k be "near" of each other. Further, by grouping the K samples into a single "global" sample of volume $N = \sum n_k$, the global mean $m = \sum n_k m_k / \sum n_k$ does not differ too much from the sample means m_k . We need a number to express how "near" the sample means m_k are from the global mean m.

The following, denoted by tradition variability between treatments, is such a number:

$$SST = \sum_k n_k (m_k - m)^2 .$$

(Initials came from *sum of squares for treatments*.)

The number SST is minimal (in fact it is 0) if and only if all sample means are equal:

$$m_1 = \ldots = m_k = \ldots = m_K \,.$$

Small values of SST appear when sample means m_k are "near" to each other, and such situations confirm the null hypothesis. When large differences between sample means do exist, at least some of them will differ considerably from the global mean, thus a large value of SST will be obtained, and this will confirm the alternative hypothesis (by rejecting the null hypothesis). However, how large should SST be to be entitled to reject the null hypothesis?

"Total" does not play any special role. Its inclusion in the results only emphasizes the fact that the statistical test is based on a decomposition of total data variance into the two variability sources: that between samples SST and that within samples.

As an example, consider the action of a drug, during 60 consecutive days, on individuals grouped into four age categories, expressed in the percentual decrease of the cholesterol level:

Under 20 years	20 – 39 years	40 – 59 years	Over 60 years
15, 17	22, 25, 20	17, 22, 28	13, 8
31, 7	36, 22, 12	15, 10	19, 16
19, 20	9, 41, 17	2, 8	22
average = 18.17	average = 2.67	average = 14.57	average = 15.60

Here $N = 27$, $K = 4$. The results offered by *Epi Info* are as follows:

ANOVA, a Parametric Test for Inequality of Population Means
(For normally distributed data only)

Variation	SS	df	MS	F statistic
Between	305.4376	3	101.8125	1.3414
Within	1745.7476	23	75.9021	
Total	2051.1852	26		

P-value = 0.2822

The reported p-value is 0.2822, thus rejection of the null hypothesis is improper (even if the discrepancy between the averages seems sufficiently large). We do not have enough data to conclude that the percentual decrease of cholesterol level depends on the age category. (Nor to conclude that it does not depend on the age category!)

4.12 Summary

To compare two or more populations, several statistical tests are available. Which one is to be used, that depends on the nature of available data from samples.

Let us present a simplified scheme:

(A) To compare two populations:

    (AA) If the data are ranked, then the Wilcoxon rank sum test is used.

    (AB) If the data are quantitative, then:

        (ABA) If one of the populations is not normally distributed, then again the Wilcoxon test is used.

        (ABB) If both populations are distributed normally, then the means μ_1, μ_2 are compared by using a t-test (which, in case of "large" samples or known variances, is in fact a z-test).

    (AC) If the data are qualitative, using a z-test compares the proportions π_1, π_2.

(B) To compare three or more populations:

    (BA) If the data are ranked, then the Kruskal-Wallis test is used.

    (BB) If the data are quantitative, then:

        (BBA) If all populations are normally distributed, then ANOVA is used.

        (BBB) If not, then again the Kruskal-Wallis test is used.

    (BC) If the data are qualitative, then a chi-square test is used.

4.13 Solved Exercises

1) A theory assesses that the number of humans possessing one of the four blood types should be proportional to

$$p^2, \; q^2 + 2p \cdot q, \; r^2 + 2p \cdot r, \text{ respectively } 2q \cdot r$$

where $p + q + r = 1$.

Given the observed frequencies 90, 180, 66 res. 49, confirm or reject the hypothesis $p = 0.4$, $q = 0.5$, $r = 0.1$.

2) A pharmaceutical company announces that drug D is 90% efficient as a painkiller for a 12-hours period.

For 160 patients out of 200, tested under strict clinical control, the drug was efficient: Is the statement of the company correct?

3) Consider the data from the paper [Doll and Pygott 1952]. Percentage changes in gastric ulcer zone, after a three-month treatment, are presented. Data from 32 admitted patients and other 32 external patients, in increasing order, are presented in the following tables:

Table 4.5. Data from admitted patients

-100	-100	-100	-100	-100	-100	-100	-100
-100	-100	-100	-100	-93	-92	-91	-91
-90	-85	-83	-81	-80	-78	-46	-40
-34	0	29	62	75	106	147	1321

Table 4.6. Data from external patients

-100	-100	-100	-100	-100	-93	-89	-80
-78	-75	-74	-72	-71	-66	-59	-41
-30	-29	-26	-20	-15	20	25	37
55	68	73	75	145	146	220	1044

Does the admission in hospital influence the results of treatment?

4) Results of a placebo-controlled clinical test trial to test the effectiveness of a sleeping drug are presented.

Patient	Drug	Placebo	Patient	Drug	Placebo
1	6.1	5.2	6	8.4	5.4
2	7.0	7.9	7	6.9	4.2
3	8.2	3.9	8	6.7	6.1
4	7.6	4.7	9	7.4	3.8
5	6.5	5.3	10	5.8	6.3

Does this evidence support the effectiveness of the drug?

5) Certain drugs differ in their side effects, depending on the gender. In a study to determine whether men or women suffer dizziness when taking a powerful drug, 8 men and 8 women were given the drug. Each was asked to evaluate the level of dizziness on a 7-point scale (1 = No effect at all – 7 = Extremely bad). The results are shown in the following table.

Men	Women	Men	Women
6	2	1	3
3	2	3	3
5	4	5	2
4	7	6	1

Can we conclude that men and women experience different levels of dizziness from the drug?

6) In the following table a comparison of birth weights (in kg) of children born to 15 non-smoker mothers with those of children born to 14 heavy smoker mothers is presented.

Non-smokers ($n = 15$)			Heavy-smokers ($n = 14$)		
3.99	3.79	3.60	3.18	2.84	2.90
3.73	3.21	3.60	3.27	3.85	3.52
4.08	3.61	3.83	3.23	2.76	3.60
3.31	4.13	3.26	3.75	3.59	3.63
3.54	3.51	2.71		2.38	2.34

Does the mother's habitude of smoking influence the birth weight of her children?

7) In 1985 the Coca-Cola Company changed the recipe of its product. Prior to the recipe change, Nielsen Company of Canada surveyed soft-drink consumer preferences and found the following percentages:

Coca-Cola   20.6%   Pepsi-Cola   18.1%

After the recipe change, another survey produced the following results:

Coca-Cola   21.4%   Pepsi-Cola   17.5%

(These figures were reported in the *Toronto Star*.) Nielsen Company of Canada samples 1000 consumers in this kind of surveys. Can we conclude that the popularity of Coca-Cola has increased after the recipe change?

Solutions.

1) We apply the chi-square goodness-of-fit test to the null hypothesis $p = 0.4$, $q = 0.5$, $r = 0.1$. Using *Excel* and organizing the computation as follows:

	A	B	C	D	E	F
1	p=	0.4				
2	q=	0.5				
3	r=	0.1				
4			Observed	Expected		
5	p^2=	0.16	90	61.6	=B5*C9	
6	q^2+2*p*q=	0.65	180	250.25	=B6*C10	
7	r^2+2*p*r=	0.09	66	34.65	=B7*C11	
8	2*q*r=	0.10	49	38.5	=B8*C12	
9				385	=SUM(D5:D8)	
10						
11				8.04E-14	=CHITEST(C5:C8,D5:D8)	

the p-value that is obtained is not large enough to support this hypothesis. (In fact, the observed data is very strong evidence against the null hypothesis!)

A more experienced statistician could put forward the values $p = 0.48$, $q = 0.36$, $r = 0.16$, for which the computed p-value 0.877 is supporting the null hypothesis.

2) The utility of the drug is confirmed in proportion $\frac{160}{200} = 0.80$ of treated patients, thus in less that 90% of cases. At first sight the statement of the company is doubtful.

However, the data used were obtained from a single sample of 200 treated patients! Perhaps the sample is an exception; data from other samples could provide results near to 90% efficiency. How large the chances for an exceptional sample are?

To answer the last question, suppose the sentence of the company is correct. Imagine all possible samples of volume 200 (extracted from the "infinite" population of treated patients) are at hand.

Denote by π the probability that drug D is efficient for a treated patient. The sentence of the company is exactly:

$$\pi = 0.90 \quad (H_0).$$

Which logical consequences of this sentence are deduced?

Denote by S the number of patients, in a sample of volume $N = 200$, for which the drug is efficient ("successful"). Let us admit S is (approximately) distributed normally. We expect the drug is efficient in

$$\mu = N \cdot \pi = 180$$

cases, and the variance of S is

$$\sigma^2 = N \cdot \pi \cdot (1 - \pi) = 18.$$

In this "population" of numbers S, the z-score of our particular sample is

$$z = \frac{160 - 180}{\sqrt{18}} \approx -4.714.$$

This score is abnormally low and corresponds to an exceptional situation. We cannot accept as true the sentence of the company.

In fact, it is easy to evaluate (in *Excel*) the probability of a number 160 or less of successful treatments with the drug:

$$P(S \leq 160) = \text{NORMDIST}(160,180,\text{SQRT}(18),\text{TRUE}) \approx 0.0000012157.$$

Notice sentences like that enounced by the pharmaceutical company can only be "rejected" when using reasoning as above, i.e. significance testing. We would like to "confirm" such sentences!

3) The average of data for admitted patients is –13.8, res. for the external patients is 15.3. At first sight there is a difference. Moreover, a paired t-test gives a p-value of 0.0066, small enough to confirm this difference, i.e. the influence of admission in hospital on the results of treatment. However, are all the conditions to apply a t-test met?

4) We use either the parametric (paired) t test, or the non-parametric Wilcoxon test. In the first case, function TTEST from *Excel* gives 0.00556 as p-value; in the second, command MEANS from *Epi Info* gives 0.0036 as p-value. Both are small enough to support the effectiveness of the drug.

5) It is a typical situation of fake quantitative data. In fact, values 1, 2, ... are qualities! The Wilcoxon rank sum test gives 0.1664 as p-value. Thus, even at the 15% significance level, dependence of gender cannot be concluded.

6) The average birth weight of children born to non-smoker mothers is 3.59, more than 12% grater than the average birth weight of children born to smoker mothers. We will accept the sentence "mother habitude of smoking diminishes the birth weight of her child" if its p-value is sufficiently small. This p-value is obtained by using a non-paired t test (heteroskedastic, in the absence of precise information). For example, in *Excel* the function TTEST will return p-value = 0.0123. (What else it is needed?)

7) The second survey reports an increase from the earlier 20.6% to 21.4% of Coca-Cola preference. However, this 0.8% increase may be casual!

Consider the null hypothesis

$$(H_0): \pi_1 = \pi_2$$

and the alternative

$$(H_1): \pi_1 < \pi_2.$$

Estimations of proportions π_1 and π_2, based on samples (of volume 1000) are $p_1 = 0.206$, $p_2 = 0.214$. The pooled proportion is

$$p = \frac{1000 \cdot p_1 + 1000 \cdot p_2}{2000} = 0.210.$$

The value of the test statistic is $z = \dfrac{p_1 - p_2}{\sqrt{p \cdot (1-p) \cdot 0.002}} \approx -0439$, and this corresponds to a p-value 0.1738.

Hence, if the significance level is chosen even at 10%, the data obtained from samples are not enough to confirm the increase in Coca-Cola preference! The increase is possibly due to hazard.

5 Bayesian (Belief) Networks

5.1 Uncertain Production Rules

Very often, knowledge of experts is presented as an ordered list of production rules. A production rule is, simply, a statement in which an IF-THEN structure is detected. Of course, production rules are stated as such by humans, which make them uncertain.

Consider a hypothesis h that may cause the apparition of the evidence e. Denote by H a Boolean[1] random variable (with only two possible values, "true" and "false"). Then h can be assimilated to the event "H is true", and by $\neg h$ we will denote the event "H is false". Similarly, E represents another Boolean random variable that helps us to describe the evidence e as the event "E is true" and, of course, $\neg e$ will represent the event "E is false". The extent of the influence of h over e is described, in probabilistic terms, as $p(e\,|\,h)$.

Suppose $p(e\,|\,h)$ is known. Of course, the evidence e may appear also when H is false ; suppose the probability $p(e\,|\,\neg h)$ is also known.

Both known probabilities above are used when stating the **uncertain complete production rule**:

IF h THEN e with probability $p(e\,|\,h)$,

　　　　　ELSE e with probability $p(e\,|\,\neg h)$.

Using general axioms of Probability Theory, the probabilities in which the "complementary evidence" $\neg e$ is involved can be easily computed:

$$p(\neg e\,|\,h) = 1 - p(e\,|\,h), \quad p(\neg e\,|\,\neg h) = 1 - p(e\,|\,\neg h).$$

It will be useful to consider all four probabilities above as the components of a matrix $\begin{pmatrix} p(e\,|\,h) & p(e\,|\,\neg h) \\ p(\neg e\,|\,h) & p(\neg e\,|\,\neg h) \end{pmatrix}$.

The uncertain production rule above is well represented diagrammatically by an arrow starting in node H and ending in node E (see Figure 5.1 below), and it is an obvious extension of the logical implication (in classical logic) $h \Rightarrow e$.

---

[1] George Boole (1815-1864), English mathematician and logician, founder of Boolean calculus.

E. Roventa and T. Spircu: Management of Knowledge Imperfection, STUDFUZZ 227, pp. 133–152.
springerlink.com   © Springer-Verlag Berlin Heidelberg 2009

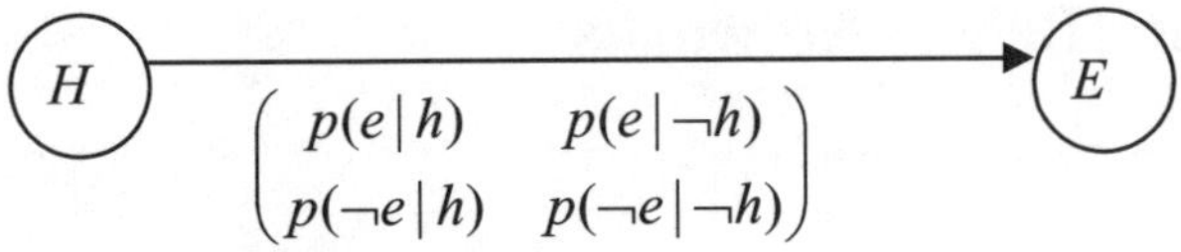

$$\begin{array}{ccc} H & \xrightarrow{\hspace{4cm}} & E \end{array}$$

$$\begin{pmatrix} p(e\,|\,h) & p(e\,|\,\neg h) \\ p(\neg e\,|\,h) & p(\neg e\,|\,\neg h) \end{pmatrix}$$

Fig. 5.1. Complete uncertain production rule

Now, the "premise" h may be also uncertain. Its uncertainty is expressed by the probability $P(h)$. Of course, automatically $P(\neg h) = 1 - P(h)$ and in fact we deal with a vector $\begin{pmatrix} P(h) \\ P(\neg h) \end{pmatrix}$ that is associated to the random variable H.

The "conclusion" e results uncertain. How we compute the number $P(e)$ that expresses its uncertainty? Of course, the computing formula should be consistent with the classical logic *Modus Ponens* reasoning rule. The corresponding vector for E, $\begin{pmatrix} P(e) \\ P(\neg e) \end{pmatrix}$, is easily obtained by the following formula:

$$\begin{pmatrix} P(e) \\ P(\neg e) \end{pmatrix} = \begin{pmatrix} p(e\,|\,h) & p(e\,|\,\neg h) \\ p(\neg e\,|\,h) & p(\neg e\,|\,\neg h) \end{pmatrix} \times \begin{pmatrix} P(h) \\ P(\neg h) \end{pmatrix}$$

that supposes the computing of a matriceal product.

In fact, what we need is only:

$$P(e) = p(e\,|\,h) \cdot P(h) + p(e\,|\,\neg h) \cdot P(\neg h).$$

However, the matriceal formula above is easily extended to variables H, E that have more than two values.

Let us consider uncertain complete production rules corresponding to logical "true" implications $a \Rightarrow b$ and $b \Rightarrow c$.

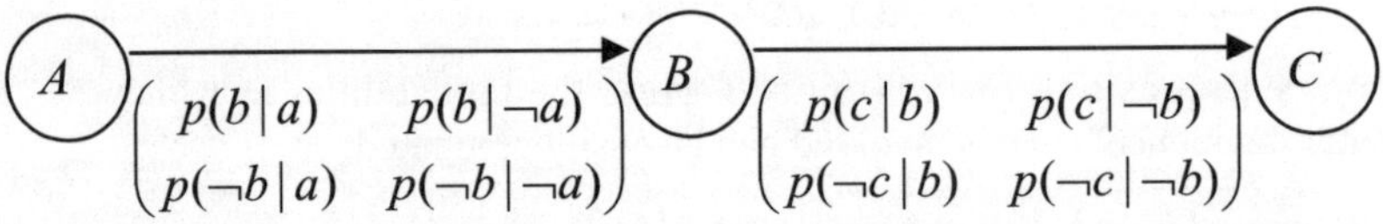

$$A \xrightarrow{\begin{pmatrix} p(b\,|\,a) & p(b\,|\,\neg a) \\ p(\neg b\,|\,a) & p(\neg b\,|\,\neg a) \end{pmatrix}} B \xrightarrow{\begin{pmatrix} p(c\,|\,b) & p(c\,|\,\neg b) \\ p(\neg c\,|\,b) & p(\neg c\,|\,\neg b) \end{pmatrix}} C$$

In classical logic, *syllogismus* allows us to infer the truth of the implication $a \Rightarrow c$. This implication corresponds to the uncertain complete production rule:

$$A \xrightarrow{\begin{pmatrix} p(c\,|\,a) & p(c\,|\,\neg a) \\ p(\neg c\,|\,a) & p(\neg c\,|\,\neg a) \end{pmatrix}} C$$

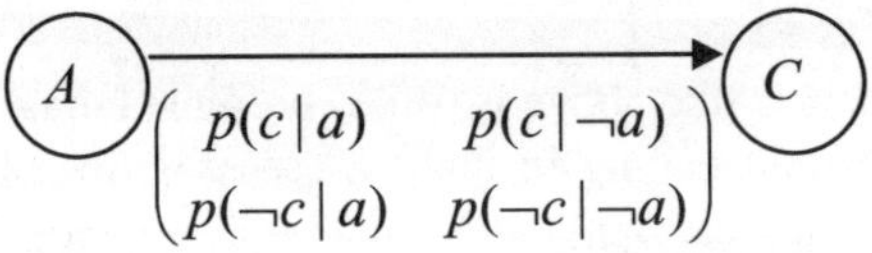

The matrix equality:

$$\begin{pmatrix} p(c|a) & p(c|\neg a) \\ p(\neg c|a) & p(\neg c|\neg a) \end{pmatrix} = \begin{pmatrix} p(c|b) & p(c|\neg b) \\ p(\neg c|b) & p(\neg c|\neg b) \end{pmatrix} \times \begin{pmatrix} p(b|a) & p(b|\neg a) \\ p(\neg b|a) & p(\neg b|\neg a) \end{pmatrix}$$

seems natural. However, it is not correct! Indeed, in general

$$p(c\,|\,a) = p(c\,|\,a \wedge b) \cdot p(b\,|\,a) + p(c\,|\,a \wedge \neg b) \cdot p(\neg b\,|\,a)$$

(see Section 3.17) instead of the supposed equality

$$p(c\,|\,a) = p(c\,|\,b) \cdot p(b\,|\,a) + p(c\,|\,\neg b) \cdot p(\neg b\,|\,a).$$

However, when

$$p(c\,|\,a \wedge b) = p(c\,|\,b) \quad \text{and} \quad p(c\,|\,a \wedge \neg b) = p(c\,|\,\neg b)$$

the computing formulas do not contradict our intuition. This happens if we consider

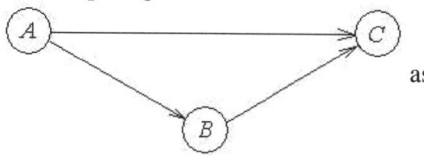

as a Bayesian network.

5.2 Bayesian (Belief, Causal) Networks

A Bayesian network is a mixed structure, combining combinatorial and probabilistic features.

Let us consider a "classical" example, due to Judea Pearl ([Pearl 1988]).

Sally's home is equipped with a burglar alarm. During the office hours Sally receives a message from John, hers neighbor, who called to say that the burglar alarm was ringing.

However, John is not "fully reliable"; he might have heard a car alarm in the street, thinks Sally.

Sally knows that the burglar alarm can be triggered by a minor earth tremor or a malfunction, and she thinks that a minor earth tremor is not affecting normally a car alarm.

Could we help Sally to evaluate quickly the chance a burglary caused John's call?

The relationship information we possess is represented in the Figure 5.2. The arrows should be interpreted as "cause" or, better, "may cause".

Five random variables: B = "burglary", T = "earth tremor", A = "alarm ring", C = "car alarm" and J = "John's call" are involved. All of them are of Bernoulli type. As before, lower-case letter x will denote the statement X = true and notation $\neg x$ will denote the statement X = false. Then b stands for "Sally's house was burgled", which is an interpretation of the statement B = true, t stands for "there was an earth tremor" and j stands for "John called". In this context, $b\,|\,j$ means "John called because Sally's house was burgled".

It is convenient to denote generically by $p(B\,|\,J)$ the set of probabilities $p(b\,|\,j)$, $p(b\,|\,\neg j)$, $p(\neg b\,|\,j)$, $p(\neg b\,|\,\neg j)$.

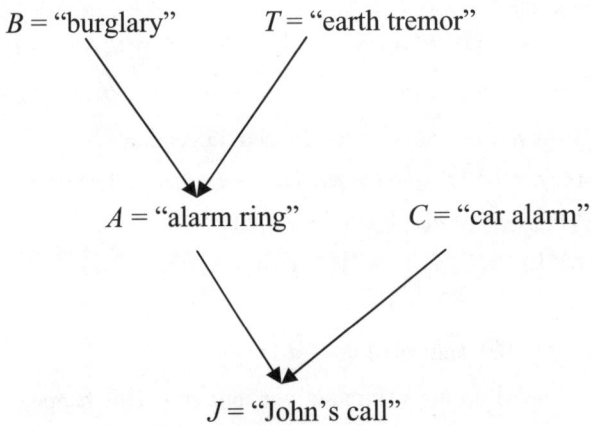

Fig. 5.2. The direct acyclic graph

We usually think that the chances a car alarm went off are not affected by whether the burglar alarm went off or the house was burgled, and Sally thinks neither an earth tremor has any influence on car alarms. That means no arrows appear between the corresponding nodes in the graph above. All this can be expressed in seven conditional independence statements:

$$p(C\,|\,A) = P(C) \qquad p(C\,|\,B) = P(C) \qquad p(C\,|\,T) = P(C)$$
$$p(C\,|\,A \wedge B) = P(C) \qquad p(C\,|\,A \wedge T) = P(C) \qquad p(C\,|\,B \wedge T) = P(C)$$
$$p(C\,|\,A \wedge B \wedge T) = P(C)$$

or in a single general statement

$$p(C\,|\,\{A, B, T\}) = P(C)\,.$$

(In fact, this statement replaces $4 + 4 + 4 + 8 + 8 + 8 + 16 = 52$ equalities between probabilities!)

Notice in the graph structure above that A, B, and T are exactly the nodes that are neither descendants of C, nor parents of C.

In general, consider a directed acyclic graph G having node set $\mathcal{N}$ and arrow set $\mathcal{A}$.

Consider a single node $V \in \mathcal{N}$. We identify

–   the subset of parents $C(V) = \{U\,|\,U \in \mathcal{N}$ and $(U,V) \in \mathcal{A}\}$,
–   the subset of descendants
    $\mathcal{D}(V) = \{W\,|\,W \in \mathcal{N}$ and there is a path from V to $W\}$,
–   the subset of other nodes $O(V) = \mathcal{N} - \{\{V\} \cup C(V) \cup \mathcal{D}(V)\}$.

A **Bayesian (belief, causal) network** is built over a directed acyclic graph $(\mathcal{N}, \mathcal{A})$. A random variable X_V is associated to each node V. A conditional table, given the variables that label the parents of V and the other nodes, is given.

More precisely, the following is satisfied in Bayesian networks:

(BN)   for any $V \in \mathcal{N}$ and $\{Z_1,...,Z_k\} \subseteq O(V)$, if $C(V) = \{U_1,...,U_s\}$, then

$$p(X_V \mid \{X_{Z_1}, X_{Z_2},..., X_{Z_k}\} \wedge X_{U_1} \wedge X_{U_2} \wedge ... \wedge X_{U_s})$$
$$= p(X_V \mid X_{U_1} \wedge X_{U_2} \wedge ... \wedge X_{U_s}).$$

(Of course, this relation could be used in multiple ways, giving to each random variable all possible values.)

For any initial node I (i.e. having no parent), a prior probability distribution, denoted by $P(I)$, should be specified.

The random variable X_V is denoted simply by V (i.e. is identified with the node) when no confusion is possible. For Bernoulli-type random variables we use lower-case letters v to denote the statement $V = $ true and notation $\neg v$ to denote the statement $V = $ false.

In general, suppose we want to obtain a conditional probability $p(v \mid w)$. Several cases have to be considered:

1) The node W is a descendent of V ($W \in \mathcal{D}(V)$). Then the first Bayes' formula

$$p(v \mid w) = \frac{p(w \mid v) \cdot P(v)}{P(w)}$$

should be used in order to reverse the role of the variables.

2) The node W is a parent of V ($W \in C(V)$). Then all the other parents U should be identified and the formula

$$p(v \mid w) = \sum_u p(v \mid u \wedge ...) \cdot p(u \wedge ... \mid w)$$

should be used.

3) The node W is neither a parent of V nor a descendant of V ($W \in O(V)$). Then we have to consider two sub-cases:

a) V has no parents. Then a particular form of the condition (BN) above is used, giving

$$p(v \mid w) = P(v).$$

b) V has parents. Then the formula

$$p(v \mid w) = \sum_u p(v \mid w \wedge u \wedge ...) \cdot p(u \wedge ... \mid w)$$

involving all the parents U of V should be used.

In the formulas above, for each parent U all its possible values u should be considered (thus u and $\neg u$ when U is Boolean).

5.3 Examples of Bayesian Networks

In scientific research we are rather often confronted with the following problem: suppose the evidence e has been detected; what are the odds of h to be the cause of the appearance of e?

In probabilistic terms, what is needed is the probability $p(h \mid e)$. Suppose all probabilities in the simplest Bayesian network (see Figure 5.1) are known. This supposes the knowledge of all values $p(e \mid h)$, $p(e \mid \neg h)$, $P(h)$. The first theorem of Bayes provides an immediate result for the probability we need:

$$p(h \mid e) = \frac{p(e \mid h) \cdot P(h)}{p(e \mid h) \cdot P(h) + p(e \mid \neg h) \cdot P(\neg h)}.$$

In the domain of medical sciences different words and notations are used. The context is as follows: a sign Σ may indicate disease Δ as a possible cause of the condition of patient. In general, if we select "at random" one individual from the population under study, then

1) He may or may not exhibit the sign. It is said that, for the sign Σ, our individual **tests positive**, res. **tests negative**.

2) The disease Δ is **present**, res. **absent** in our individual.

Therefore, in the context above, for each individual one and only one of the following four different possibilities may appear:

a) Tests positive $(\Sigma +)$ and disease is present $(\Delta +)$. This is the situation of **true positive** (TP) individuals.

b) Tests positive $(\Sigma +)$ but disease is absent $(\Delta -)$. This situation is encountered in **false positive** (FP) individuals.

c) Tests negative $(\Sigma -)$ but disease is present $(\Delta +)$. The individuals in this category are **false negative** (FN).

d) Tests negative $(\Sigma -)$ and disease is absent $(\Delta -)$. This is the situation of **true negative** (TN) individuals.

In the language of the theory of probability, the **sensitivity** of a sign Σ – of course, for the disease Δ – is defined as the probability that an individual "X" tests positive, provided the disease is present. Thus,

$$Sensitivity = p("X" \text{ is } \Sigma + |"X" \text{ has } \Delta)$$
$$= p("X" \text{ is } \Sigma + |"X" \text{ is TP} \vee "X" \text{ is FN}).$$

This probability is estimated by the ratio $\dfrac{f_{TP}}{f_{TP} + f_{FN}}$ whenever data obtained from a "good" sample from the population is available. (Of course, f_{TP} is the absolute frequency of the "true positive" individuals in the sample, etc.)

The sensitivity is a measure of accuracy of predicting "positive events". Usually a sign Σ is considered relevant for the disease Δ if the value of the sensitivity is "large", i.e. over 0.75 (75%). Of course, values over 0.9 (90%) or, better, over 0.95 (95%) are to be preferred.

The **specificity** of a sign Σ – for the disease Δ – is defined as the probability that an individual tests negative, provided the disease is absent. Thus,

$$Specificity = p("X" \text{ is } \Sigma- | "X" \text{ has not } \Delta)$$
$$= p("X" \text{ is } \Sigma- | "X" \text{ is FP} \vee "X" \text{ is TN}).$$

This value is estimated by the ratio $\dfrac{f_{TN}}{f_{FP} + f_{TN}}$.

Notice that the specificity is a measure of error in predicting "negative events". High values of the specificity, the best larger than 0.95 (95%) are preferred.

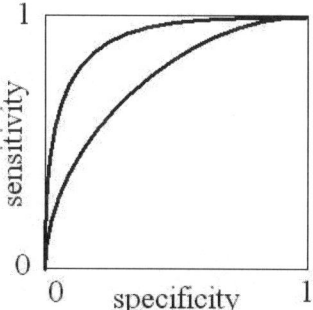

Fig. 5.3. Two ROC curves

The **receiver operating characteristic** curve (in short, the ROC curve) is constructed by plotting the sensitivity versus the specificity. It is larger for a model with higher predictive accuracy (see Figure 5.3).

The area under a ROC curve is called the **c-statistics** and varies from a minimum of 0.5 (when the predictions are as good as "pure chance") to a maximum of 1.

Another very important notion in our context (of a sign Σ and a disease Δ) is the **predictive value** of the sign for the respective disease. This is defined as the probability that the disease is present, provided the individual tests positive, i.e. by the formula:

$$Predictive\, value = p("X" \text{ has } \Delta | "X" \text{ is } \Sigma+).$$

This value is estimated by the ratio $\dfrac{f_{TP}}{f_{TP} + f_{FP}}$. However, a more interesting computing formula, involving the notions presented above, can be established from the first theorem of Bayes:

$$Predictive\ value = \frac{Sensitivity \cdot Prevalence}{Sensitivity \cdot Prevalence + (1 - Specificity) \cdot (1 - Prevalence)}\ .$$

Here the **prevalence** of the disease Δ is simply the probability that the disease is present in an individual of the population, and is estimated by the incidence of the disease in a sample, i.e. by the ratio

$$\frac{f_{TP} + f_{FN}}{f_{TP} + f_{FN} + f_{FP} + f_{TN}}\ .$$

The prevalence of a disease is "small", usually under 1%. The predictive value of a test is "much larger", usually around 10%.

A very simple example of Bayesian network (with only three nodes H, E_1, E_2) is presented in Figure 5.4.

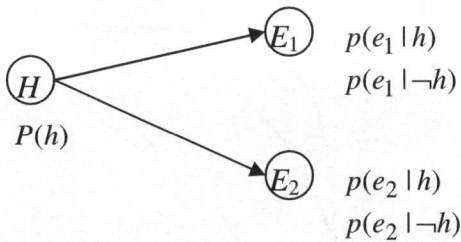

Fig. 5.4. Multiple evidence caused by a hypothesis?

This network corresponds, obviously, to two IF-THEN-ELSE rules:

IF h THEN e_1 with prob. $p(e_1 \mid h)$ ELSE e_1 with prob. $p(e_1 \mid \neg h)$

IF h THEN e_2 with prob. $p(e_2 \mid h)$ ELSE e_2 with prob. $p(e_2 \mid \neg h)$.

The relations (BN), for the nodes, are as follows

$$p(E_1 \mid \{E_2\}, H) = p(E_1 \mid H),$$
$$p(E_2 \mid \{E_1\}, H) = p(E_2 \mid H).$$

Suppose we detected both "evidences" e_1 and e_2 occurred (i.e. E_1 and E_2 are both true). What is the probability that h occurred as well?

We need to compute $p(h \mid e_1 \wedge e_2)$. Because both E_1 and E_2 are descendents of H, we use the Bayes' formula

$$p(h \mid e_1 \wedge e_2) = \frac{p(e_1 \wedge e_2 \mid h) \cdot P(h)}{p(e_1 \wedge e_2 \mid h) \cdot P(h) + p(e_1 \wedge e_2 \mid \neg h) \cdot P(\neg h)}\ .$$

Now, in order to compute the two probabilities needed in this formula, we use axiom (P3) – see Section 3.6 – then a relation (BN)

$$p(e_1 \wedge e_2 \mid h) = p(e_1 \mid h) \cdot p(e_2 \mid e_1 \wedge h) = p(e_1 \mid h) \cdot p(e_2 \mid h)$$

and analogously

$$p(e_1 \wedge e_2 \mid \neg h) = p(e_1 \mid \neg h) \cdot p(e_2 \mid \neg h).$$

This is nothing else that expressions of conditional independence of "events" e_1 and e_2. Thus, in this simple case, relation (BN) expresses the conditional independence of "evidences".

An interesting observation: if the event "E_2 is true" is, under the condition "H is true", more probable that under the condition "H is false", i.e. if

$$p(e_2 \mid h) \geq p(e_2 \mid \neg h)$$

then

$$p(h \mid e_1 \wedge e_2) \geq p(h \mid e_1);$$

hence the probability of "H is true" increases if the evidence e_2 occurred after the occurrence of evidence e_1. On the contrary, if $p(e_2 \mid h) < p(e_2 \mid \neg h)$, it follows $p(h \mid e_1 \wedge e_2) < p(h \mid e_1)$, in other words the probability of the hypothesis h decreases.

Consider another simple case of Bayesian network, having also three nodes H_1, H_2 and E.

The relations (BN) are non-trivial only for two nodes:

$$p(H_1 \mid \{H_2\}) = P(H_1),$$

$$p(H_2 \mid \{H_1\}) = P(H_2)$$

and these relations express in fact the conditional independence of random variables H_1, H_2 (representing hypotheses that could "cause" the evidence).

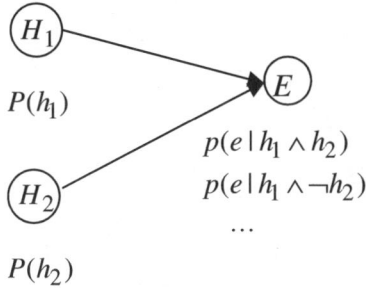

Fig. 5.5. Independent hypotheses

Suppose we detected e (i.e. E is true) occurred. What is the probability that h_1 occurred as well, but not h_2? The answer is simple, and is obtained by computing $p(h_1 \wedge \neg h_2 \mid e)$.

Node E is a descendent of both H_1 and H_2; therefore, we use the first Bayes' formula

$$p(h_1 \wedge \neg h_2 \mid e) = \frac{p(e \mid h_1 \wedge \neg h_2) \cdot P(h_1 \wedge \neg h_2)}{P(e)}.$$

Using the independence of H_1 and H_2, we obtain

$$P(h_1 \wedge \neg h_2) = P(h_1) \cdot P(\neg h_2).$$

Now, the parents of E are exactly H_1 and H_2, hence

$$\begin{aligned}
P(e) &= p(e \mid h_1 \wedge h_2) \cdot P(h_1 \wedge h_2) + p(e \mid h_1 \wedge \neg h_2) \cdot P(h_1 \wedge \neg h_2) \\
&+ p(e \mid \neg h_1 \wedge h_2) \cdot P(\neg h_1 \wedge h_2) + p(e \mid \neg h_1 \wedge \neg h_2) \cdot P(\neg h_1 \wedge \neg h_2) \\
&= p(e \mid h_1 \wedge h_2) \cdot P(h_1) \cdot P(h_2) + p(e \mid h_1 \wedge \neg h_2) \cdot P(h_1) \cdot P(\neg h_2) \\
&+ p(e \mid \neg h_1 \wedge h_2) \cdot P(\neg h_1) \cdot P(h_2) + p(e \mid \neg h_1 \wedge \neg h_2) \cdot P(\neg h_1) \cdot P(\neg h_2).
\end{aligned}$$

Consider the possible causes of a patient's aching hands or elbows. These causes could be: tennis elbow, arthritis, or dishpan hands. In the Figure 5.6 the acyclic graph of relations is represented.

It can be seen that aching hands does not directly depend on whether the patient has an aching elbow or tennis elbow. An aching elbow depends only on whether the patient has tennis elbow or arthritis. Whether patients have tennis elbow depends on whether they have arthritis. In addition, presenting dishpan hands does not directly depend on any of the other variables.

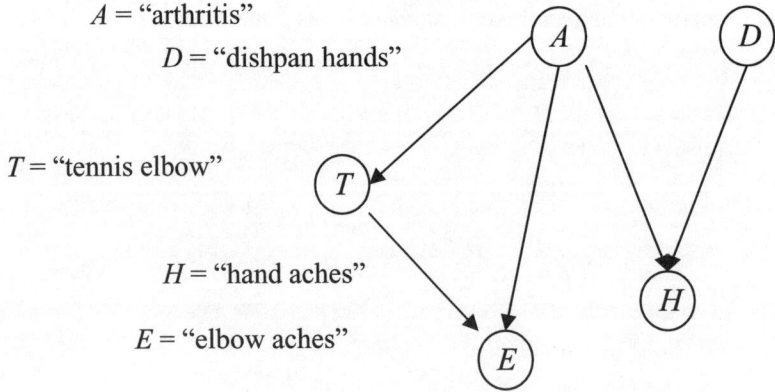

A = "arthritis"

D = "dishpan hands"

T = "tennis elbow"

H = "hand aches"

E = "elbow aches"

Fig. 5.6. Example of a DAG

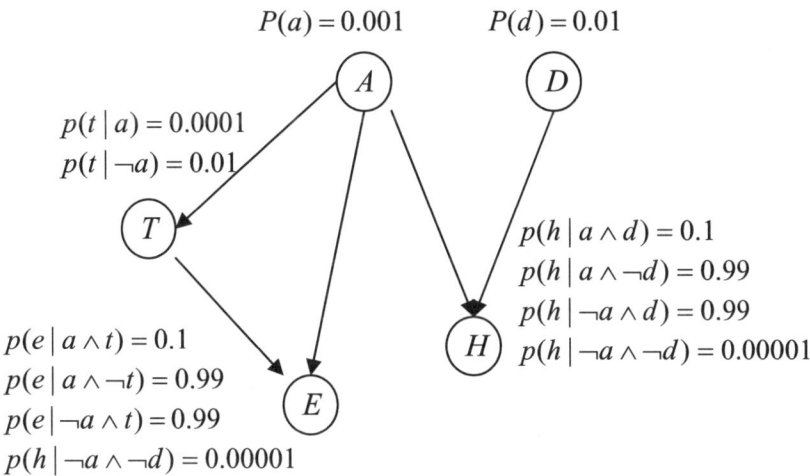

$P(a) = 0.001$ $P(d) = 0.01$

$p(t \mid a) = 0.0001$
$p(t \mid \neg a) = 0.01$

$p(h \mid a \wedge d) = 0.1$
$p(h \mid a \wedge \neg d) = 0.99$
$p(h \mid \neg a \wedge d) = 0.99$
$p(h \mid \neg a \wedge \neg d) = 0.00001$

$p(e \mid a \wedge t) = 0.1$
$p(e \mid a \wedge \neg t) = 0.99$
$p(e \mid \neg a \wedge t) = 0.99$
$p(h \mid \neg a \wedge \neg d) = 0.00001$

Fig. 5.7. Example of a belief network

In order to present this situation as a belief network we have to specify two prior probabilities and several conditional probabilities. Let us present all these data in the Figure 5.7 above.

It can be seen from the data that arthritis and dishpan hands, separately, determine almost sure (0.99) hand aches; however, a patient with both has a very low chance (0.1) to suffer from hand aches.

Conditions (BN) exist for all five nodes. They are as follows:

$p(A \mid \{D\}) = P(A)$,
$p(D \mid \{A, E, T\}) = P(D)$,
$p(E \mid \{D, H\} \wedge A \wedge T) = p(E \mid A \wedge T)$,
$p(H \mid \{E, T\} \wedge A \wedge D) = p(H \mid A \wedge D)$,
$p(T \mid \{D, H\} \wedge A) = p(T \mid A)$.

Let us compute the probability that our patient has tennis elbow, in the absence of any other information:

$$P(t) = p(t \mid a) \cdot P(a) + p(t \mid \neg a) \cdot P(\neg a)$$
$$= 0.0001 \cdot 0.001 + 0.01 \cdot (1 - 0.001) = 0.0099901 \, (\approx 1\%)$$

Once aching elbow is observed,

$$p(t \mid e) = \frac{p(e \mid t) \cdot P(t)}{P(e)}$$

$$= \frac{p(e \mid t) \cdot P(t)}{p(e \mid t) \cdot P(t) + p(e \mid \neg t) \cdot P(\neg t)} = 0.908935 \, (91\%),$$

thus tennis elbow becomes "almost sure".

In the same manner,

$$p(a \mid e) = ... = 0.0909 \,(9\%),$$

raises approx. 90 times, and

$$p(d \mid e) = P(d) = 0.01,$$

because of a condition (BN).

Suppose both aching elbow and aching hands are observed; then

$$p(t \mid e \wedge h) = 0.0917 \,(9\%)$$

and the chances of a tennis elbow drop approx. 10 times,

$$p(a \mid e \wedge h) = 0.908 \,(91\%),$$

thus arthritis becomes "almost sure", and

$$p(d \mid e \wedge h) = 0.0926 \,(9\%),$$

thus the chances of dishpan hands raise 9 times.

In general, in a Bayesian network any conditional probability can be found algorithmically, taking into account the graph structure and the natural ordering of nodes, as exemplified above.

5.4 Software

Heavy computing is necessary even with simple Bayesian networks. Therefore, programmed software is needed.

Netica, developed by Norsys Software Corporation, is probably the most widely used software in this category. As the developers say, it is "simple, reliable, and high performing ... it is the tool of choice for many of the world's leading companies and government agencies".

The following Bayesian network – see Figure 5.8 – from [Lauritzen and Spiegelhalter 1988] is presented as a teaching example in **Netica**.

The arrows represent, as usual, possible cause-effect links or influences. The two top nodes represent "hypotheses"

VisitAsia – a possible visit to Asia, *Smoking* – smoking habitude
which may influence the likelihood of

Tuberculosis – tuberculosis disease, *Cancer* – lung cancer disease,
Bronchitis – bronchitis disease
and eventually of the two bottom nodes ("evidences" or symptoms for diseases)

XRay – X-Ray result, *Dispnea* – dispnea.

The intermediate node *TbOrCa* represents "tuberculosis or lung cancer". Obviously, the variable associated to this node depends, in a deterministic way, from the random variables associated to *Tuberculosis* and *Cancer*. Therefore, this node is

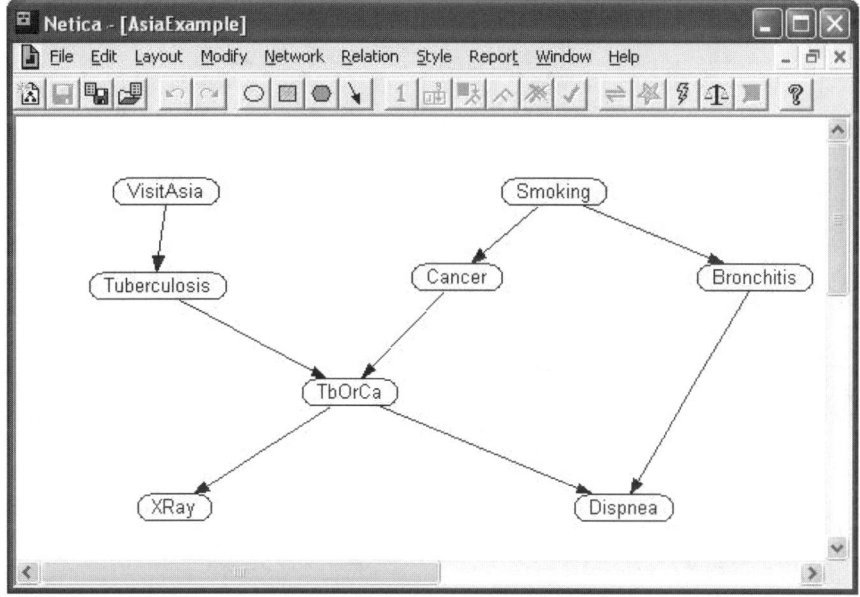

Fig. 5.8. *Netica* – an example (I)

"deterministic", and this is clearly stated in its description inside the file **AsiaExample.dne**:

```
node TbOrCa {
  kind = NATURE;
  discrete = TRUE;
  chance = DETERMIN;
  states = (true, false);
  parents = (Tuberculosis, Cancer);
  functable =                // Tuberculosis  Cancer
          ((true,            // present       present
            true),           // present       absent
           (true,            // absent        present
            false));         // absent        absent  ;
  equation = "TbOrCa (Tuberculosis,Cancer) = Tuberculosis||Cancer";
};
```

The Figure 5.9 below presents the results *Netica* produced for a heavy smoker whose X-Ray is abnormal and who visited Asia within last 3 years. The chances of lung cancer are 53%.

Of course, the results are strongly dependent on the conditional probabilities "declared" in the file. An excerpt of the description for the node *Dispnea* is as follows:

```
node Dispnea {
  kind = NATURE;
  discrete = TRUE;
  chance = CHANCE;
  states = (present, absent);
  parents = (TbOrCa, Bronchitis);
  probs =
```

```
      // present        absent         // TbOrCa  Bronchitis
      (((0.72,          0.28),         // true     present
        (0.65,          0.35)),        // true     absent
       ((0.15,          0.85),         // false    present
        (0.01,          0.99)));       // false    absent      ;
};
```

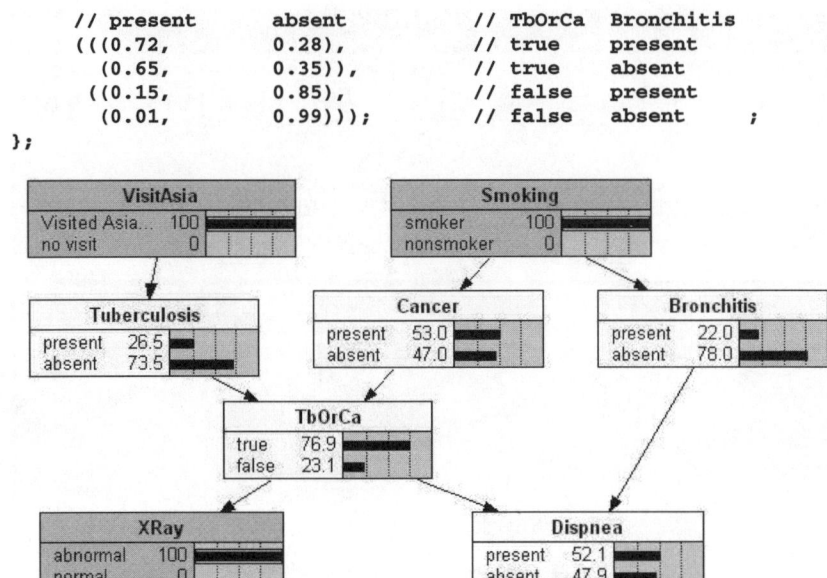

Fig. 5.9. *Netica* – an example (II)

Netica is a powerful tool. It is able to deal with general discrete random variables (having several "states") and also with continuous variables. However, in the latter case, only a finite number of "states" are dealt with, obtained by declaring separation levels.

BayesBuilder is a simpler tool, developed by SNN group at the University of Nijmegen, The Netherlands. It treats data stored in bbnet files, as for example **asia.bbnet** (see Figure 5.10).

In such files, data about nodes are dispersed in three groups, separating the states, the parents, and the role. For example, the data about "Dispnea" node appears as follows:

```
node node2
{ type = table
  name = "DYSPNOEA"
  desc = "Dyspnoea"
  states = {"TRUE", "FALSE"}
  info = ""
  refs { }
}

prob node2
{ parents = {node7, node1} //TBORCA, BRONCHITIS
  p = {  // TRUE FALSE
           0.72  0.28 // TRUE TRUE
           0.65  0.35 // TRUE FALSE
           0.15  0.85 // FALSE TRUE
           0.01  0.99 // FALSE FALSE
       }
}
role node2
{ use "10" // TRUE FALSE
  test "dispnea"
}
```

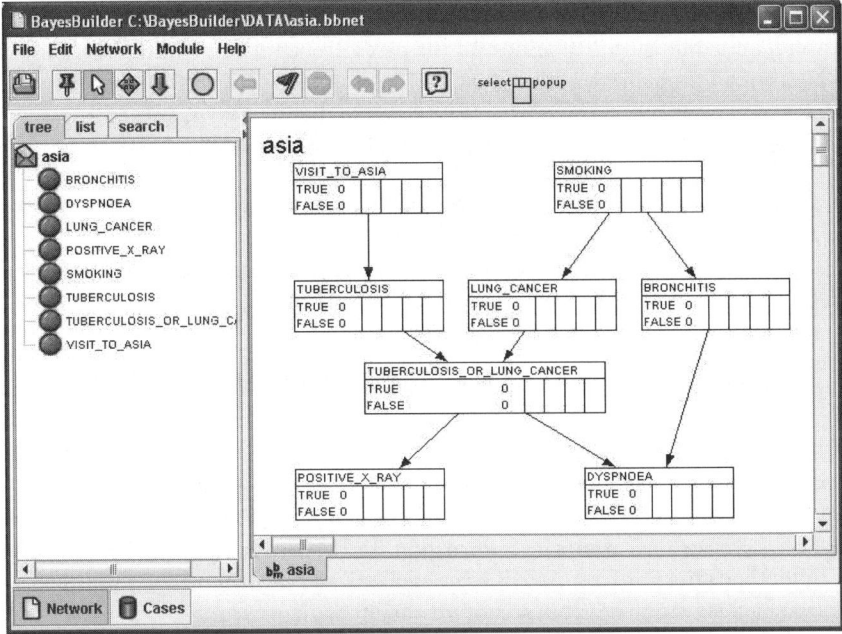

Fig. 5.10. *BayesBuilder* – same example (I) as in Fig. 5.8

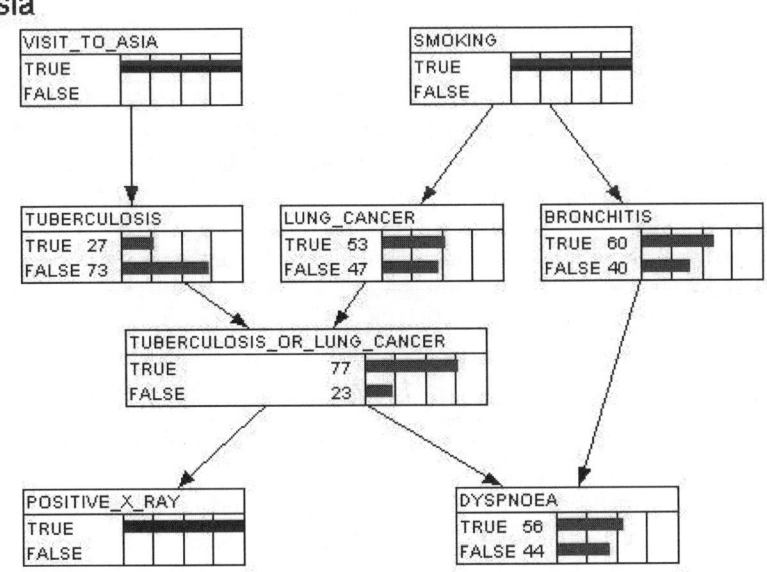

Fig. 5.11. *BayesBuilder* – same example (II) as in Fig. 5.9

Results presented in Figure 5.11 differ from those presented in Figure 5.9 because of different conditional probabilities for the pair (Bronchitis, Smoking). Only the first two digits of all probabilities are shown.

However, a very interesting example is stored in the file **alarm.bbnet**.

Charles River Analytics Inc. develops the BNet family, which contains:

a) *BNet.Builder* software, to create belief networks, entering probabilistic information, and rapidly getting results, and

b) *BNet.EngineKit*, which is software developed to incorporate Bayesian networks technology into other applications.

5.5 Bias of the Bayesian (Probabilistic) Method

The knowledge (about effects and their probable causes) is embodied in Bayesian network in a more compact form that in rule-based expert systems.

Of course, the conditional probabilities "inserted" in a Bayesian network are estimated on the basis of "historical" information, and on this basis it can be used to predict future behavior – of course, in a probabilistic way. Such a tool – we refer to a well-designed Bayesian network – is of great help in diagnosis.

However, Bayesian reasoning is based on three assumptions:

1) Hypotheses are mutually exclusive,
2) Hypotheses are exhaustive,
3) Conditional independence of evidences under both hypothesis and its negation.

Often these conditions are not met. The probabilistic approach supposes also to estimate *a priori* all the probabilities. Often this is difficult.

Restrictions concerning the values of probabilities should be taken into account; for example, the sum of probabilities of all hypotheses has to be exactly 1.

The computation of all intermediate values is mandatory. However, the majority of these coefficients have a minor influence on the final result. Thus, inefficacity is present. And the total ignorance is never taken into account.

5.6 Solved Exercises

1) An admissions committee of a college is trying to determine the probability that an admitted applicant is really qualified.

As a rule, qualified people have high grade point average. However, only around 90% of qualified people are able to obtain excellent recommendations.

About a half of non-qualified people also possess excellent recommendations and about a quarter have high grade point average.

The admissions committee "admits" all applicants who have high grade point average and possess excellent recommendations. Of course, all applicants who have not a high grade point average and do not possess excellent recommendations are "rejected". The admissions committee "admits" half of the other applicants.

Which probability the admission committee is trying to estimate?

2) ([Rich and Knight 1991]) Consider the following set of propositions:
 patient has spots,
 patient has measles,
 patient has high fever,
 patient has Rocky Mountain Spotted Fever,
 patient has previously been inoculated against measles,
 patient was recently bitten by a tick,
 patient has an allergy.

a) Create a network that defines the causal connections among the corresponding nodes.

b) Make it a Bayesian network by constructing the necessary conditional probabilities.

3) (A typical Judea Pearl exercise.) The belief network shown below formalizes the following situation: you have a new burglar alarm installed at home. It is fairly reliable at detecting a burglary, but also responds on occasion to minor earthquakes. You also have two neighbors, John and Mary, who have promised to call you at work when they hear the alarm. John quite reliably calls when he hears the alarm, but some times confuses the telephone ringing with the alarm and calls then too. Mary, on the other hand, likes rather loud music and sometimes misses the alarm altogether.

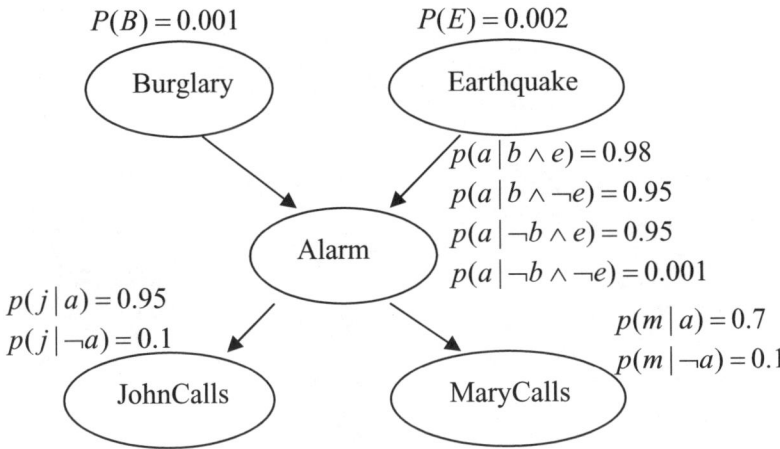

Compute the joint probability that neither John nor Mary calls and that there is both an earthquake and a burglary. That is, compute $P(\neg j \wedge \neg m \wedge b \wedge e)$.

Solutions. 1) The relevant probabilities are given in the Bayesian network shown below.

Here: a means "applicant is qualified";

b means "applicant has high grade point average",

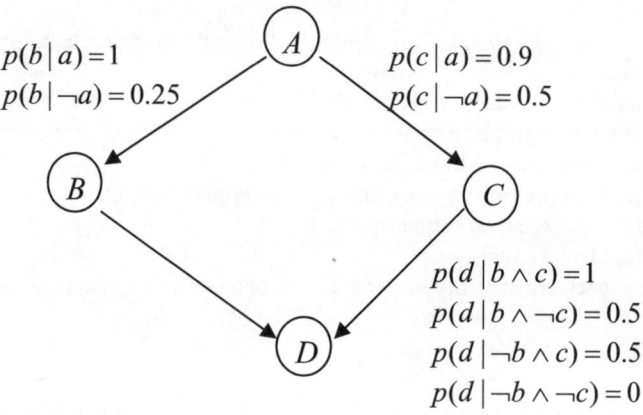

$p(b\,|\,a) = 1$
$p(b\,|\,\neg a) = 0.25$

$p(c\,|\,a) = 0.9$
$p(c\,|\,\neg a) = 0.5$

$p(d\,|\,b \wedge c) = 1$
$p(d\,|\,b \wedge \neg c) = 0.5$
$p(d\,|\,\neg b \wedge c) = 0.5$
$p(d\,|\,\neg b \wedge \neg c) = 0$

c means "applicant is in possession of excellent recommendations",

d means "applicant is admitted".

The commission needs to compute $p(a\,|\,d)$.

However, a supplementary parameter needs to be specified. Namely, one needs to estimate the "initial" probability of qualified people. If this is $P(a) = 0.3$, then $p(a\,|\,d) = 0.52$ (see the figure below). If $P(a) = 0.6$, then $p(a\,|\,d) = 0.79$.

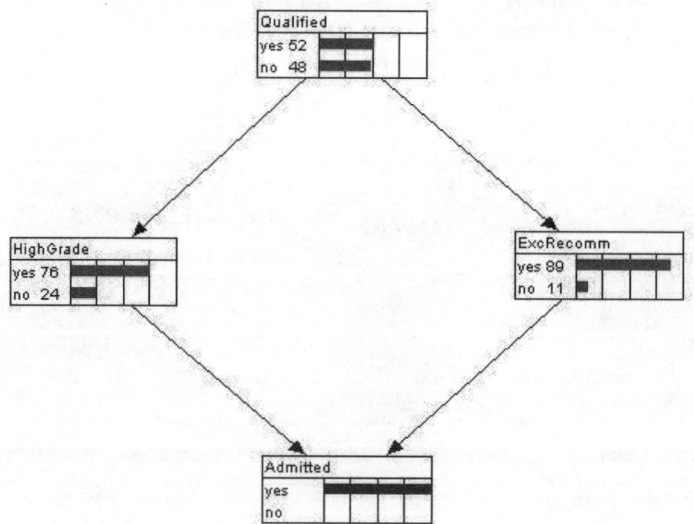

2) a) A possible global description of the causal connections between the propositions is contained in the following directed acyclic graph:

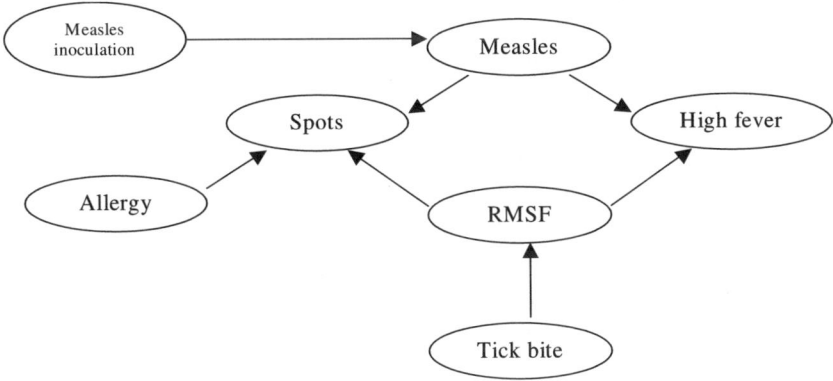

b) The original example in [Rich and Knight 1991] is based on the following probabilities:

	Probability
P(inoculation)	0.4
P(tickbite)	0.0001
P(allergy)	0.1
p(measles|inoculation)	0.000001
p(measles|¬inoculation)	0.0001
p(RMSF|tickbite)	0.1
p(RMSF|¬tickbite)	0
p(fever|RMSF∧¬measles)	0.9
p(fever|RMSF∧measles)	0.91
p(fever|¬RMSF∧¬measles)	0.001
p(fever|¬RMSF∧measles)	0.8
p(spots|allergy∧RMSF∧¬measles)	0.85
p(spots|allergy∧RMSF∧measles)	0.95
p(spots|allergy∧¬RMSF∧¬measles)	0.1
p(spots|allergy∧¬RMSF∧measles)	0.85
p(spots|¬allergy∧RMSF∧¬measles)	0.8
p(spots|¬allergy∧RMSF∧measles)	0.9
p(spots|¬allergy∧¬RMSF∧¬measles)	0.001
p(spots|¬allergy∧¬RMSF∧measles)	0.8

Of course, different values are allowed.

Thus, the probability of an allergy is only 10%. However, if our patient has spots, then with 91% probability our patient has an allergy.

If, furthermore, the patient has high fever, then the probability of an allergy diminishes to 26% and there is 13% probability that patient was recently bitten by a tick.

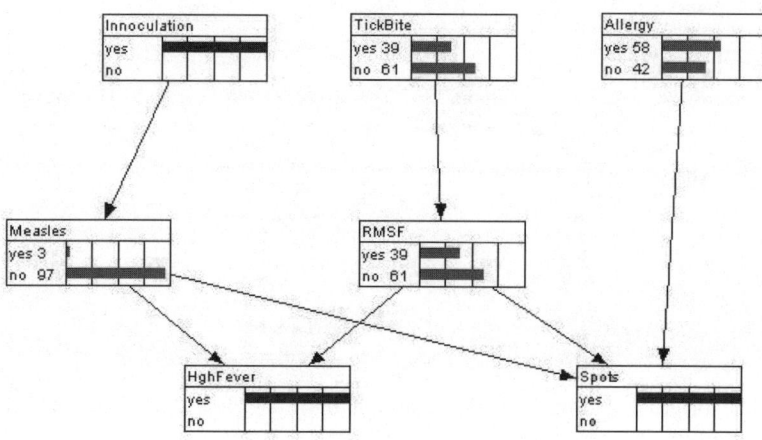

Now, if we know that the patient has previously been inoculated against measles, then the probability of an allergy rises to 58% and the probability that patient was recently bitten by a tick rises to 39%.

6 Certainty Factors Theory

6.1 Certainty Factors

Bayes' formulas are complex enough and definitely not adequate to human's brain reasoning activities. Certainty factors theory is an alternative to Bayesian reasoning – when reliable statistical information is not available or the independence of evidence cannot be assumed – and introduces a certainty factors calculus based on the human expert heuristics.

It is well known that the human experts express their estimation by using terms such as "most probable", "probably", "probably not", and "improbable". However, these terms do not have the same meaning as in Probability Theory!

Uncertainty is expressed usually in a linguistic manner, by using words such as "usually", "frequently", and „sometimes". It is usual to attach to uncertain knowledge a numerical certainty factor. There are several transformation tables of linguistic terms in numerical ones. [Negnevitski 2000] provides us two – in which the attached values are between 0 and 100 – obtained after investigation of college students (which cannot be catalogued, yet, as experts)

Term	After [Simpson 1944]	After [Hakel 1968]
always	99	100
usually	85	79
...		
almost never	3	2

The study lead by [Nakao and Axelrod 1983] has shown that there is a difference in perception between physicians and other type of experts, and also between experts and laymen about how to estimate uncertainty values (induced by differences in education and tradition). Some examples:

Term	Experts		[Negnevitski 2000]
	Physicians	Non-physicians	
often	40	59	78 res. 74
frequently	50	45	73 res. 72
not often	25	28	13 res. 16

E. Roventa and T. Spircu: Management of Knowledge Imperfection, STUDFUZZ 227, pp. 153–160.
springerlink.com   © Springer-Verlag Berlin Heidelberg 2009

It is to notice that the certainty factor of (the truth of) any evidence e, as evaluated by a human expert, will have value in the interval $[-1, 1]$. The value -1 means "definitely not (true)". The value 0 means "maximal uncertainty" or "total ignorance". For the medical "facts" only positive values are used.

On the other hand, in reasoning, we quite frequently use rules that are not "certain". These rules have also a certainty factor. The certainty factor of the rule will be combined with the certainty factor of premises to obtain an evaluation of the certainty factor of conclusion.

Psychological experiments have shown that in medical reasoning knowledge is responsible for the most of uncertainty (and not the reasoning itself).

6.2 Stanford Algebra

When MYCIN was created (1974), the developers (of Stanford University – see [Shortliffe, Davis, Axline, Buchanan, Green and Cohen 1975]) decided to implement in its inference engine several special rules to manage uncertainty. These are known today as Stanford Algebra.

The name MYCIN comes from a common suffix of several anti-microbial agents, and denotes a classical expert system for the diagnosis therapy of blood infections and meningitis.

In MYCIN, the process of selection of a therapy is decomposed in four parts:

1) The infection needs a treatment?
2) If yes, identification of the organisms susceptible to be responsible is performed, then;
3) Medication to be recommended is selected, and;
4) Treatment is prescribed.

The original version of MYCIN includes more than 200 production rules, all of the same type. Most of them are heuristics, based on facts.

Facts about the "world" are represented as possible uncertain sentences

> parameter (context) predicate value (certainty factor).

Examples:

> site (of the culture) is blood (c.f. = 1), i.e. absolutely certain,
> identity (of the organism) is Klebsiella (c.f. = 0.8), i.e. almost certain,
> identity (of the organism) is Proteus (c.f. = –0.6), i.e. probably not.

Here is an example of a production rule in MYCIN – rule 85 – expressed in plain English:

> if   the site of the culture is blood,
>   the Gram stain of the organism is negative,
>   the morphology of the organism is rod, and
>   the patient is a compromised host,
> then   there is a suggestive evidence (0.6) that the identity of the organism is Pseudomonas æruginosa.

The number 0.6 is assimilated to the certainty factor of the rule.

For each production rule a **certainty factor** is defined, which represents the level of belief in the conclusion of the rule given the evidence contained in the antecedent (considered as precise).

An important part of the inference engine of MYCIN is composed by strategy rules (which are meta-rules). The meta-rules permit to reduce or to reorder the set of rules to be used. These rules are also heuristic, so imprecise. Here is an example:

> if   the patient is a compromised host,
>   there exist rules mentioning Pseudomonas in premises, and
>   there exist rules mentioning Klebsiella in premises,
> then   it is suggested (0.4) to use the rules mentioning Pseudomonas first.

The certainty factor of a composed premise A is computed – starting from the certainty degrees of the components – according to the formulas:

$$cf(A) = \max\{cf(A_1), cf(A_2)\} \text{ if } A = A_1 \vee A_2,$$
$$cf(A) = \min\{cf(A_1), cf(A_2)\} \text{ if } A = A_1 \wedge A_2,$$
$$cf(A) = -cf(A') \text{ if } A = \neg A'.$$

Before applying an uncertain production rule "IF A, THEN B (cf)" the certainty factor of the premise A is computed. If negative, the rule will be ignored. If positive, the certainty factor of the conclusion B will be computed by multiplying the rule's certainty factor with the certainty factor of the premise A.

Hence

$$cf(\text{conclusion}) = cf(\text{premise}) \cdot cf(\text{rule}).$$

Example. Consider the rule 85 above. We know for sure, before applying this rule, that the site of the culture is blood ($cf = 1$), and that the Gram stain of the organism is negative ($cf = 1$). We appreciate that the morphology of the organism is rod ($cf = 0.9$) and that the patient is a compromised host with the certainty factor 0.8.

Since the premise of the rule is a conjunction, the certainty factor of the premise is taken as a minimum, so 0.8. From here, we obtain by multiplication, the certainty factor $0.8 \cdot 0.6 = 0.48$ of the conclusion, "it is Pseudomonas".

If the same conclusion B is obtained by applying another rule, then its certainty factor is reinforced by the following formula:

$$cf(B) = \begin{cases} cf_{\text{old}} + cf_{\text{new}} - cf_{\text{old}} \cdot cf_{\text{new}} & \text{if } cf_{\text{old}} \text{ and } cf_{\text{new}} \text{ are positive} \\ cf_{\text{old}} + cf_{\text{new}} + cf_{\text{old}} \cdot cf_{\text{new}} & \text{if } cf_{\text{old}} \text{ and } cf_{\text{new}} \text{ are negative} \\ \dfrac{cf_{\text{old}} + cf_{\text{new}}}{1 - \min\{|cf_{\text{old}}|, |cf_{\text{new}}|\}} & \text{in all other situations.} \end{cases}$$

The formula above corresponds to an operation (with real numbers in the interval $[-1, 1]$)

$$\alpha * \beta = \begin{cases} \alpha + \beta - \alpha \cdot \beta & \text{if } \alpha, \beta \geq 0 \\ \alpha + \beta + \alpha \cdot \beta & \text{if } \alpha, \beta \leq 0, \\ \dfrac{\alpha + \beta}{1 - \min\{|\alpha|, |\beta|\}} & \text{if } \alpha \cdot \beta < 0 \end{cases}$$

which is commutative and associative. These proprieties of operation $*$ make sure the independence of final certainty factor of a conclusion, no matter the order in which the rules are treated.

In the expert system MYCIN, the numbers cf are used in the heuristic research to establish a priority for the proposed goals. The methods for combining cf's are derived from probabilities and are related to probabilities. However, they are distinctly different [Shortliffe and Buchanan 1975].

6.3   Certainty Factors and Measures of Belief and Disbelief

Consider a production rule

IF e THEN h

having the certainty factor cf. (Here cf represents the belief in hypothesis h given that evidence e has occurred.)

If the evidence e is uncertain, and its certainty factor was evaluated at $cf(e)$, then the certainty factor $cf(h \mid e)$ of the hypothesis h (based on e) will be computed as follows:

$$cf(h \mid e) = cf(e) \cdot cf .$$

From here, the certainty factor of the hypothesis h is adjusted as follows

$$cf_{new}(h) = cf_{old}(h) \cdot cf(h \mid e).$$

Example. Suppose two rules have as conclusion the same hypothesis, the two certainty factors were obtained (from human experts):

Rule 1: IF e_1 THEN h ($cf_1 = 0.5$)
Rule 2: IF e_2 THEN h ($cf_2 = 0.25$)

we know for sure the evidence e_1 (i.e. $cf(e_1) = 1$), we are almost sure of evidence e_2 (i.e. $cf(e_2) = 0.8$), and we are totally ignorant about h.

Starting from $cf_0(h) = 0$, after the first rule is fired we obtain

$$cf_1(h) = 0 * cf(h \mid e_1) = cf(h \mid e_1) = cf(e_1) \cdot cf_1 = 0.5 .$$

Then, after the second rule is fired,

$$cf_{12}(h) = cf_1(h) * cf(h \mid e_2) = 0.5 * (cf(e_2) \cdot cf_2)$$
$$= 0.5 * (0.8 \cdot 0.25) = 0.5 + 0.2 - 0.5 \cdot 0.2 = 0.6 \, .$$

Suppose we know for sure another evidence e_3 and a third rule having conclusion h exists:

Rule 3: IF e_3 THEN h ($cf_3 = -0.4$)

After this rule is fired,

$$cf_{123}(h) = cf_{12}(h) * cf(h \mid e_3) = 0.6 * (-0.4 \cdot 1) = \frac{0.6 - 0.4}{1 - 0.4} = 0.333$$

and our confidence in hypothesis h is drastically reduced.

Now, how a certainty factor of a rule is obtained? Given a rule

IF e THEN h,

the degree of confidence in the hypothesis h (due to evidence e), i.e the certainty factor $cf(h \mid e)$, was originally defined in MYCIN as a difference

$$MB(h \mid e) - MD(h \mid e)$$

where MB is a measure of belief in the hypothesis h due to e, and MD is a measure of disbelief in h due to e.

Thus, in MYCIN a simple way to combine belief and disbelief into a single number was chosen.

The measure of belief was defined in terms of probabilities as follows:

$$MB(h \mid e) = \begin{cases} 1 & \text{if } P(h) = 1 \\ \dfrac{\max\{p(h \mid e), P(h)\} - P(h)}{1 - P(h)} & \text{otherwise} \end{cases}$$

and the measure of disbelief as follows:

$$MD(h \mid e) = \begin{cases} 1 & \text{if } P(h) = 0 \\ \dfrac{P(h) - \min\{p(h \mid e), P(h)\}}{P(h)} & \text{otherwise.} \end{cases}$$

Thus, when lacking evidence we have $p(h \mid e) = P(h)$ and, if h is not "absolutely certain" or "impossible", then $MB = MD = 0$, hence $cf = 0$.

On the other side, $cf = 0$ could appear also when $MB = MD > 0$. This means the belief is cancelled out by the disbelief.

In general, cf is a number between -1 and 1. A positive value of the cf means the evidence supports the hypothesis, since $MB > MD$.

As it was specified above, the numbers cf are used in MYCIN to rank different hypotheses.

However, there are difficulties with this approach because a single piece of disconfirming evidence could drastically cancel all the confirmation of many previous pieces of evidence. For example, nine pieces of evidence could produce a measure of belief of 0.9 and a tenth piece of evidence could come with $MD = 0.8$, thus on the whole $cf = 0.1$.

In addition, in order to activate a rule in MYCIN, the premise of the rule should have a certainty factor of at least 0.2. (This threshold value 0.2 was chosen in order to minimize the activation of rules that only weakly suggest a hypothesis, thus to increase the overall system efficiency.)

In a later version of MYCIN the definition of the certainty factor was changed into the following:

$$cf = \frac{MB - MD}{1 - \min\{MB, MD\}}$$

to soften the effect of a single disconfirming piece of evidence on many confirming pieces of evidence.

Using this definition, from the data in the example above we obtain

$$cf = \frac{0.9 - 0.8}{1 - 0.8} = \frac{0.1}{0.2} = 0.5$$

thus a value over the activation threshold!

The major advantage of certainty factors was the simple computations by which uncertainty propagates in the system. Moreover, the certainty factors are easy to understand and clearly separate belief from disbelief.

However, there are difficulties with the theoretical foundations of certainty factors. One problem is that the cf values could rank hypotheses in opposite order as conditional probabilities do (and some people do not accept this behavior).

A second "strange" problem is the transitivity formula

$$cf(h|e) = cf(h|i) \cdot cf(i|e)$$

valid for an intermediate hypothesis i based on evidence e. Notice that in Probability Theory, in general

$$p(h|e) \neq p(h|i) \cdot p(i|e).$$

The success of MYCIN – despite all these problems – is probably due to short inference chains and simple hypotheses. For situations that are more complex serious reasoning errors could be obtained. However, the importance of this type of expert system consists mainly in the quality of its production rules.

6.4 Solved Exercises

1) The teacher noticed – and all the students agreed upon – that last month there were twice as many dry days as rainy days.

One of the students noticed that there were no two consecutive rainy days. Hence, he added, the certainty factor attached to

IF today_is_dry THEN tomorrow_is_rain

should be set to 0.5. Explain and correct the mistake.

2) From two different human experts we obtained the following:

  IF today_is_dry AND temperature_is_warm
     THEN tomorrow_is_rain (it's possible enough cf = 0.4)
  IF today_is_dry AND sky_is_overcast
     THEN tomorrow_is_rain (most probably cf = 0.7)

Today is a dry and warm day, and the sky is overcast. What can be said about the weather forecast?

3) Given the following uncertain rules in a reasoning system

$$A \wedge (\neg B) \Rightarrow C \ (0.9)$$
$$C \vee D \Rightarrow E \ (0.75)$$
$$F \Rightarrow A \ (0.8)$$
$$G \Rightarrow D \ (0.8)$$

and the uncertain facts

$$B \ (-0.8)$$
$$F \ (\alpha)$$
$$G \ (\beta)$$

use Stanford algebra to determine the certainty factor of E.

4) All the rules of the expert system in Chapter 1, Exercise 3 have $\alpha > 0$ as certainty factors. Knowing facts A and B are sure (I.e. their c.f. = 1), find the certainty factors of C and D.

5) Verify that the operation $*$ is associative (i.e. satisfies

$$\alpha * (\beta * \gamma) = (\alpha * \beta) * \gamma$$

for all $\alpha, \beta, \gamma \in [-1, 1]$).

Solutions. 1) We know from 30 days only 10 were rainy, the other 20 dry. Now, after a dry day in 10 cases (50%) followed another dry day, and in the other 10 cases a rainy day. If this pattern (tendency?) will continue, then we are totally uncertain about what will follow after a dry day. That means the certainty factor should be set to 0, not to 0.5. Certainty actors are not probabilities!

It is sure that after a rainy day a dry day followed. Therefore we are entitled to set to 1 the certainty factor attached to

IF today_is_rain THEN tomorrow_is_dry

(despite the fact that the weather pattern is subject to change).

Without the observation of the student, the distribution of the 10 rainy days among the other 20 dry days could vary. Each human being, according to his/hers past experience, sets the certainty factor of the rule above. Remember, −1 means "zero chances to be true".

2) The aggregate certainty factor has a value $0.4 * 0.7 = 0.82$. Thus tomorrow_is_rain is almost sure!

3) For $\alpha \in [-1, 1]$ we have $0.8 \cdot \alpha \le 0.8$, hence C has certainty factor $0.9 \cdot 0.8\alpha$. On the other hand, D has certainty factor 0.8β. Now, the certainty degree of E is $0.75 \cdot \max\{0.72\alpha, 0.8\beta\}$.

4) We have $\min\{\alpha * \alpha^2, \alpha\} = \alpha$, and the respective certainty factors are $\alpha * \alpha^2$ and $\alpha * \alpha^2 * \alpha^3$.

5) When comparing $\alpha * (\beta * \gamma) = (\alpha * \beta) * \gamma$ for $\alpha, \beta > 0$ and $\gamma < 0$, consider the following three cases: a) $\beta \ge -\gamma$; b) $\beta < -\gamma$ and $\alpha \ge -\beta * \gamma$; c) $\beta < -\gamma$ and $\alpha < -\beta * \gamma$.

7 Belief Theory

7.1 Belief Approach

The following example is extracted from [Degoulet and Fieschi 1999].

Suppose a decision agent (physician) faces the following problem: a 68-years-old diabetic man, who injured his left foot, has developed an infection that may cause gangrene.

Two therapeutic solutions are possible:

- To amputate immediately, or
- To treat with anti-inflammatory medication and wait.

The first solution may cause death. The second solution may cure the infection or, if the medication is ineffective, may require a larger amputation (above the knee) or even cause death.

This is a typical decision situation. A decision tree (see Figure 7.1) can represent it.

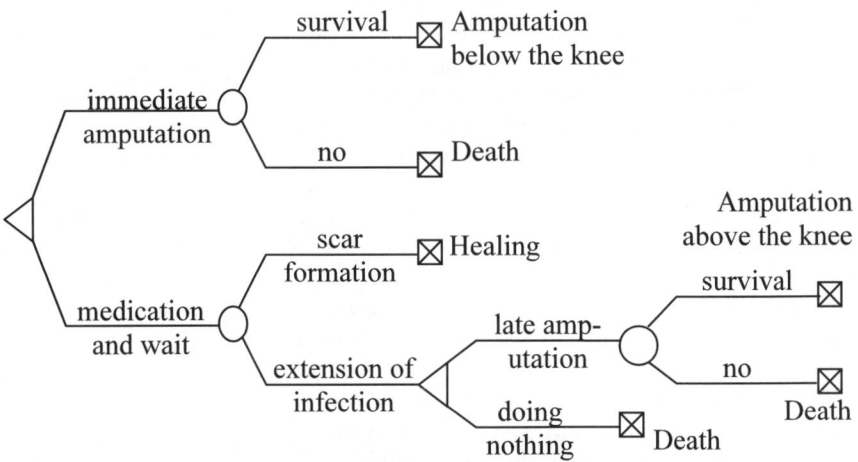

Fig. 7.1. Example of a decision tree

E. Roventa and T. Spircu: Management of Knowledge Imperfection, STUDFUZZ 227, pp. 161–186.
springerlink.com   © Springer-Verlag Berlin Heidelberg 2009

Of course, the decision of our agent depends on several parameters:

- The likelihood of death during surgery,
- The likelihood of the extension of infection, and
- The usefulness (utility) of each final result.

A "computed" utility is associated to any possible decision, and it seems obvious that the "rational" decision is that which corresponds to the maximal utility. However, in computing utilities the first two of the above parameters play a decisive role, and these are of "probabilistic" type. The agent's belief, based on his/hers knowledge, is involved and needs to be estimated. Of course, his/hers personal experience plays a major role in estimating "beliefs".

As pointed out in [Smets 2000], beliefs manifest themselves at two mental levels:

1) The *credal level*, before the particular decision is made, when beliefs are entertained and updated,
2) The *pignistic level*[1], when beliefs are converted into probabilities used to make decisions.

The motivation for the use of probabilities is usually linked to "rational" behavior in making decisions (see [Savage 1954]). However, this motivation is justified only at the pignistic level. At the credal level, "beliefs" are not bound to satisfy the classical conditions imposed on probabilities.

By convention, "the probability of a proposition is 0" means that our proposition is assumed to be definitely false, i.e. no new evidence will alter our belief. The probability of a proposition reflects the agent's ignorance about the truth of that proposition. However, agents observe the world(s), and they modify/update their beliefs based on new "knowledge" acquired.

Beliefs in propositions can be "measured" also in terms of numbers between 0 and 1.

Before presenting the main ideas of Belief Theory, let us remind some ideas from Probability Theory that support this theory.

The frequentist approach to treat probabilities has an important advantage: it is supported by mathematical analysis, i.e. by limit calculus, and the computations are relatively simple. However, the main weakness is serious: it applies only to inherently repeatable events; the probability of a future singular event is undefined. And events in biology, economics, sociology and politics are unique!

The major advantage of the Bayesian approach is its flexibility: at least in principle each uncertain proposition can be treated. However, neither can be guaranteed that we will maintain subjective evaluations – judged today as "correct" – unchanged in the future, nor that other people will accept our evaluations as "correct". The Bayesian approach is highly subjective!

We notice here that very few physicians, and the vast majority of ordinary people, are aware of the computations imposed by the "correct rules" of Bayesian reasoning. In fact they are not following, by the book, the rules of Probability Theory. In majority they over-estimate, more or less drastically, probabilities. However, due to their personal experience, physicians overcome this tendency of over-estimating probabilities and usually prescribe adequate treatment to patients.

---

[1] In Latin *credo* = to believe, *pignus* = bet.

There is no objective way to assess the status of a living patient, or to identify the optimal therapeutic plan. Based on his personal previous experience, the physician treating this patient is reaching an opinion about the status – i.e. is forming a belief – and is prescribing a treatment on the basis of this belief.

Typically such a belief is expressed as one or more sentences like this:

"the patient x has a disease d from group g"

or, formally,

```
patient(x) ∧ disease(d) ∧ group(g)
            ∧ is_from(d,g) ∧ has(x,d).
```

Of course, the belief is highly subjective and depends strongly on the education and experience of our physician.

In fact, the belief should be expressed as one or more sentences like this:

"the physician p believes the patient x has disease d from group g"

or, formally,

```
physician(p)∧patient(x)∧disease(d)∧group(g)
            ∧is_from(d,g)∧believes(p,has(x,d),b)
```

where b denotes the degree of belief.

Notice that apparently there is no uncertainty in the last sentence, though the uncertainty of

```
patient(x) ∧ disease(d) ∧ group(g)
            ∧ is_from(d,g) ∧ has(x,d)
```

in fact of

```
has(x,d)
```

is expressed by the evaluator-agent p and it is b. It is obvious that degree of belief b depends on physician p.

7.2 Agreement Measures

How could we measure the agreement between two (or more) subjective beliefs (about the state of the same patient, at a given "moment" in time)? How could we fusion two different beliefs?

Let us underline this term: agreement. In case numbers express the opinions of the "domain experts" or "raters", the term "similarity" would be appropriate. However, in our case the belief of physicians is expressed in categorical terms, this is why we prefer the term "agreement".

A large number of agreement measures have been suggested in the past and can be found in the literature (see [Hripcsak and all 2002]). Let us review two classical ones: the observed agreement and the kappa coefficient.

In two categories (+ + and − − below) both raters agree and in the other two disagree.

Table 7.1.

		Rater R_1 judgement	
		Positive	Negative
Rater R_2 judgement	Positive	(Category + +) a cases	(Category – +) c cases
	Negative	(Category + –) b cases	(Category – –) d cases

The observed agreement between R_1 and R_2 is computed by the formula

$$OA = \frac{a+d}{a+b+c+d}$$

It is obvious that the values of OA are rational numbers between 0 and 1. Values of OA near 1 express a high degree of agreement between our raters; on the contrary, values of OA near 0 express a low degree of agreement.

Table 7.2

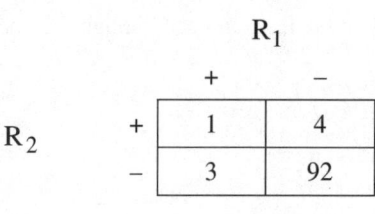

However, how should the "middle value" ½ be interpreted? Accepting the opposition high-low, the value ½ represents a medium degree of agreement. However, in our minds a "medium degree of agreement" does not coincide with "a threshold between agreement and disagreement"!

Indeed, we perceive the degree of agreement as a degree of belief in the sentence "the two raters agree on their judgments" and from here the confusion can be immediate! The following probabilistic analysis will reveal the point. If nothing is known about our raters, except that both are "guessing" at random, then the prior probability of a positive evaluation by a rater is ½, and the probability of a case to be classified in any of the four categories is ¼; hence the expected agreement is ½. In such way, a "medium degree of agreement" is obtained purely by chance!

The serious drawback of this agreement measure OA appears when a is relatively small compared to b and c, and very small compared to d. Unfortunately, this is the situation (of interest!) of rare diseases, where the vast majority of cases fall in Category – –. The value OA in this case is very large (i.e. near to 1), whatever the other opinions are! (See the three examples in tables 7.2, 7.3, 7.4: for each the degree of agreement is 0.93, even when no agreement on positive cases was detected!

Another agreement measure is the **kappa coefficient**, first introduced in [Cohen 1960]. To define the kappa coefficient, one has to understand first how the expected agreement EA is defined. Its definition is analogous to that of OA:

$$EA = \frac{E(a) + E(d)}{E(a+b+c+d)}$$

where $E(\bullet)$ denotes the expected value of a random variable.

If a number of cases were distributed in the four categories above (see Table 7.1), proportional to the opinions of both raters about the observed values, then we expect to find, in Category $+\,+$, around $\dfrac{(a+b)\cdot(a+c)}{a+b+c+d}$ cases. In fact,

$$E(a) = \frac{(a+b)\cdot(a+c)}{a+b+c+d}$$ and, analo-

gously, $E(d) = \dfrac{(b+d)\cdot(c+d)}{a+b+c+d}$.

Hence, the formula of the expected agreement is

$$EA = \frac{(a+b)\cdot(a+c)+(b+d)\cdot(c+d)}{(a+b+c+d)^2}$$

Table 7.3.

		R_1	
		+	−
R_2	+	0	4
	−	3	93

Table 7.4.

		R_1	
		+	−
R_2	+	0	0
	−	7	93

Now, the formula of the kappa coefficient $\kappa = \dfrac{OA - EA}{1 - EA}$ is set is such a way that chance agreement becomes 0.

A direct formula is immediate:

$$\kappa = \frac{2(a\cdot d - b\cdot c)}{(a+b)\cdot(a+c)+(b+d)\cdot(c+d)} .$$

The values of κ are rational numbers between -1 and 1.

7.3 Dempster–Shafer Theory

Let us confine ourselves to the creedal level only and explore how beliefs of individual agents are entertained or modified.

In the previous chapters we explored two possibilities to treat uncertainty in a system of rules of the form

IF e THEN h

namely:

- The probabilistic approach (statistical, Bayesian, and logical) which assume the estimation of *a priori* probability of h (i.e. the *a priori* probability that hypothesis H is true), and likelihood of the rule;
- The approach using certainty factors *cf*.

In [Dempster 1967] and then in [Shafer 1976] another approach has been proposed, based on belief and plausibility measures.

The initial point of Dempster–Shafer approach is to consider agents that possess knowledge encoded in their evidential corpus. Based on his/hers knowledge, each agent *You* "believes" something about the "real world" ω_0, which is one of the possible worlds (elements of the universe Ω) detected or imagined by *You*. Changes into his/hers "system of beliefs" are possibly induced by supplementary pieces of evidence e "known" by our agent.

More precisely, given a subset A of Ω, the belief of our agent in the truth of the sentence "the real world ω_0 is in A", expressed as a number between 0 and 1, is denoted by Bel(A). Our agent *You* entertains a system of beliefs $\{\text{Bel}(A) \mid A \subseteq \Omega\}$ that is fully determined, at each moment, by the knowledge possessed at that moment.

Thus the system of beliefs held by our agent *You* (at a moment t) is nothing else than a function

$$\text{Bel} : 2^\Omega \to [0,1].$$

To better understand the conditions such a function satisfies it is convenient to identify sentences "the real world ω_0 is in A" with the corresponding subsets A of Ω. Of course, if the above propositions are interpreted as subsets of Ω, then the disjunction res. conjunction of two propositions is interpreted as the union res. intersection, and the implication is interpreted as inclusion.

Let us normalize beliefs. More precisely, consider the agent *You* possesses a unit of belief and denote by $m(A)$ that part of the whole belief of *You* that supports A and, due to lack of information, does not support any strict subset of A. ("Supports A" means "believes that the real world ω_0 is in A".) Of course, the number $m(\emptyset)$ is expressing the degree of confusion existent in the beliefs of agent *You*.

A **normalized basic belief assignment** (see [Smets 2000]) is a function $m : 2^\Omega \to [0,1]$ satisfying two conditions:

(BBA1) $m(\emptyset) = 0$ (i.e. "no confusion at all"),

(BBA2) $\sum_{A \subseteq \Omega} m(A) = 1$ (i.e. "exhaustion of all belief possessed by *You*").

A general basic belief assignment is bound to satisfy only (BBA2), i.e. it allows the existence of a certain amount of confusion in the beliefs of *You*.

We suppose that the sum appearing in (BBA2) is defined in some precise way. In practical applications the universe Ω is finite, and in this case the definition is clear.

It is a fundamental difference between this definition and that of a probability measure for which we asked that

$$\sum_{\omega \in \Omega} m(\omega) = 1.$$

Therefore, the probability measure is defined on Ω while the basic belief assignment is defined on 2^Ω.

For example, suppose that universe Ω has only three elements. In the Probability Theory, a probability measure defined on Ω is expressed by a random variable X with three values. Particularly we have the situation induced by a (perfect) die tossed for which $\Omega = \{1, 2 \text{ or } 3, 4 \text{ or } 5 \text{ or } 6\}$, and

$$m(1) = \frac{1}{6}, \ m(2 \vee 3) = \frac{2}{6}, \ m(4 \vee 5 \vee 6) = \frac{3}{6}.$$

In the Dempster–Shafer theory we can consider that Ω is composed of the elementary propositions a, b, c. Then 2^{Ω} contains also their negations $\neg a = b \vee c$, $\neg b = a \vee c$, $\neg c = a \vee b$ and two exceptional subsets, $\varnothing$ and Ω itself. The basic belief assignment m expresses the degree of belief in each possible proposition, elementary or not. Recall for $A \subseteq \Omega$ $m(A)$ is interpreted as that part of the belief of an agent in the sentence "the actual world belongs to A" and not in all sentences "the actual world belongs to a particular strict subset of A". Of course, $m(\Omega)$ may be non-zero; it expresses the global uncertainty still existing in the belief system. For example, the following table describes a particular normalized basic belief assignment:

subsets A	$\{a\}$	$\{b\}$	$\{c\}$	$\{a,c\}$	$\{b,c\}$	Ω
measure $m(A)$	0.1	0.3	0	0.3	0.2	0.1

We notice in this table that:

a) $m(\Omega) = 0.1$, thus it is not required that $m(\Omega) = 1$;

b) $\{b\} \subset \{b,c\}$ (i.e. $b \Rightarrow b \vee c$), and $m(\{b\}) > m(\{b,c\})$, thus it is not required that $m(A) \leq m(B)$ when $A \subseteq B \subseteq \Omega$;

c) No relationship between the measure of $\{a\}$ and the measure of its complement $\{b,c\}$ is required;

d) There exist propositions, such as c, that have zero measure. Generally, the propositions p for which $m(p) > 0$ are called **focal**. The name comes from the fact that we focalize our interest on these propositions;

e) There is a subset, $\{a,b\}$, that does not appear in the table. The reason is simple: it is not supported at all, its measure is zero.

Recall that the propositions a of universe Ω are interpreted as subsets A of the set Ω. The empty subset $\varnothing$ symbolizes the contradiction. A "true" implication $b \Rightarrow a$ is interpreted as an inclusion $B \subseteq A$.

If A is a given arbitrary subset (i.e. a is a proposition) of universe Ω, then generally we distinguish three types of subsets (propositions) of Ω:

1) The subsets B (i.e. propositions b) that are contained in A (thus for which $b \Rightarrow a$ is true);

2) The subsets C (i.e. propositions c) that are contained in the complement of A (thus for which $c \Rightarrow \neg a$ is true);

3) The parts D (i.e. propositions d) that intersect both A and its complement $\Omega - A$ (thus for which both $d \Rightarrow a$ and $d \Rightarrow \neg a$ are not true).

By analogy with the probability definition of an event, we define the **belief** of a subset A (i.e. of a proposition a) of universe Ω, using only the subsets (propositions) of type 1, by the formula

$$\text{Bel}(A) = \sum_{B \subseteq A} m(B) \quad \text{(alternate definition } \text{Bel}(a) = \sum_{b \Rightarrow a} m(b)\text{)}.$$

In such way we obtain the **belief measure**

$$\text{Bel}: 2^{\Omega} \to [0, 1].$$

Condition (BBA2) is interpreted as $\text{Bel}(\Omega) = 1$.

The subsets (propositions) of type 2 of A are used – in the same manner as above – to define the belief of the negation ("propositional complement") $\text{Bel}(\neg a)$.

Because of the existence of parts of type 3, it is obvious that generally,

$\text{Bel}(a) + \text{Bel}(\neg a) \leq 1$ for each proposition a.

The subsets of type 3 are used to define the degree of uncertainty between a and its negation $\neg a$. However, the subsets of type 1 and 3 are used together, to define the **plausibility** of subset A (i.e. of proposition a), by the formula

$$\text{Pl}(A) = \sum_{B \cap A \neq \emptyset} m(B) \quad \text{(alternate definition } \text{Pl}(a) = \sum_{\neg(b \Rightarrow a)} m(b)\text{)}.$$

It is immediate that

$$\text{Pl}(a) + \text{Bel}(\neg a) = 1,$$

thus $\text{Pl}(a) = 1 - \text{Bel}(\neg a)$.

It results directly from the definitions that, in general,

$$\text{Bel}(a) \leq \text{Pl}(a)$$

and that the difference $\text{Pl}(a) - \text{Bel}(a)$, i.e. the interval $[\text{Bel}(a), \text{Pl}(a)] \subseteq [0, 1]$, "measures" the uncertainty of proposition a.

Consider two propositions a_1 and a_2 such that $a_1 \Rightarrow a_2$. If $b \Rightarrow a_1$, it is clear that $b \Rightarrow a_2$ because of syllogism rule. Then,

$$\text{Bel}(a_1) = \sum_{b \Rightarrow a_1} m(b) \leq \sum_{b \Rightarrow a_2} m(b) = \text{Bel}(a_2)$$

therefore, the belief measure satisfies the monotonic condition

(Bel1)   If $a_1 \Rightarrow a_2$, then $\text{Bel}(a_1) \leq \text{Bel}(a_2)$.

We know that if $a_1 \Rightarrow a_2$, then $\neg a_2 \Rightarrow \neg a_1$. Thus we deduce that $\text{Bel}(\neg a_2) \leq \text{Bel}(\neg a_1)$, hence $\text{Pl}(a_2) \geq \text{Pl}(a_1)$. The plausibility measure satisfies the monotonic condition:

(Pl1)   If $a_1 \Rightarrow a_2$, then $\text{Pl}(a_1) \leq \text{Pl}(a_2)$.

Consider two propositions a_1 and a_2, and their disjunction $a_1 \vee a_2$. To obtain

$$\text{Bel}(a_1 \vee a_2) = \sum_{b \Rightarrow a_1 \vee a_2} m(b)$$

it is necessary to evaluate all "subsets" B of $A_1 \cup A_2$. There are six possibilities (see Figure 7.2).

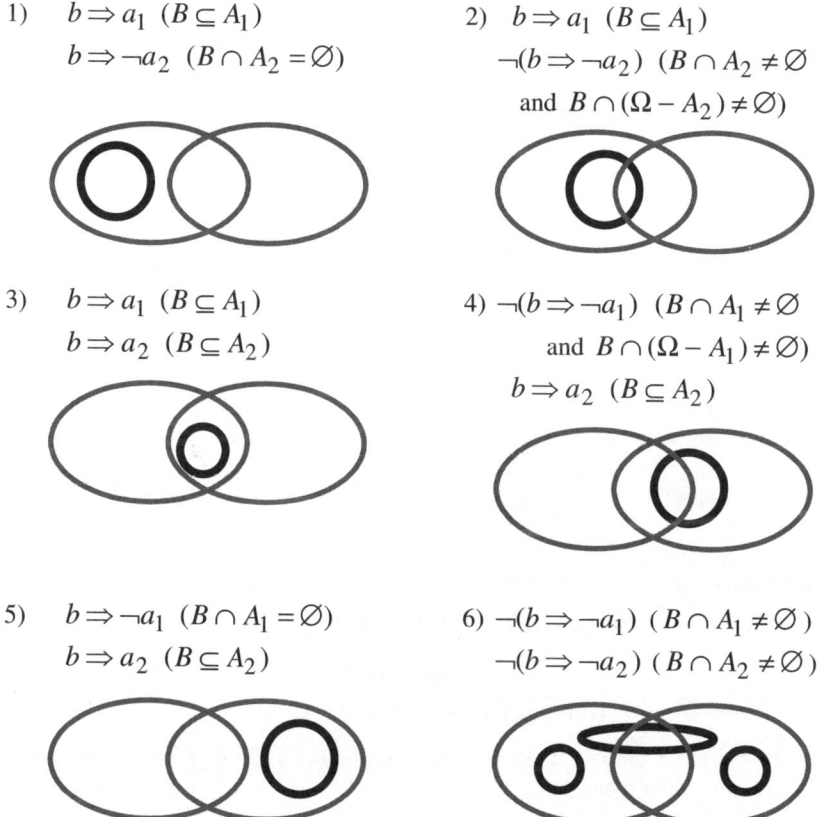

1) $b \Rightarrow a_1$ $(B \subseteq A_1)$
   $b \Rightarrow \neg a_2$ $(B \cap A_2 = \emptyset)$

2) $b \Rightarrow a_1$ $(B \subseteq A_1)$
   $\neg(b \Rightarrow \neg a_2)$ $(B \cap A_2 \neq \emptyset$
   and $B \cap (\Omega - A_2) \neq \emptyset)$

3) $b \Rightarrow a_1$ $(B \subseteq A_1)$
   $b \Rightarrow a_2$ $(B \subseteq A_2)$

4) $\neg(b \Rightarrow \neg a_1)$ $(B \cap A_1 \neq \emptyset$
   and $B \cap (\Omega - A_1) \neq \emptyset)$
   $b \Rightarrow a_2$ $(B \subseteq A_2)$

5) $b \Rightarrow \neg a_1$ $(B \cap A_1 = \emptyset)$
   $b \Rightarrow a_2$ $(B \subseteq A_2)$

6) $\neg(b \Rightarrow \neg a_1)$ $(B \cap A_1 \neq \emptyset)$
   $\neg(b \Rightarrow \neg a_2)$ $(B \cap A_2 \neq \emptyset)$

Fig. 7.2. Subsets of a union

From situations 1-2-3 we retrieve $b \Rightarrow a_1$, and from 3-4-5 we retrieve $b \Rightarrow a_2$. In addition, it is obvious that 3 means $b \Rightarrow a_1 \wedge a_2$. Thus we can write:

$$\text{Bel}(a_1 \vee a_2) = \text{Bel}(a_1) + \text{Bel}(a_2) - \text{Bel}(a_1 \wedge a_2) + \sum m(b) \tag{6}$$

and from here we deduce the sub-additivity of a belief measure:

(Bel2) $\text{Bel}(a_1 \vee a_2) \geq \text{Bel}(a_1) + \text{Bel}(a_2) - \text{Bel}(a_1 \wedge a_2)$.

By contrast, the plausibility measure is super-additive:

(Pl2) $\text{Pl}(a_1 \vee a_2) \leq \text{Pl}(a_1) + \text{Pl}(a_2) - \text{Pl}(a_1 \wedge a_2)$.

In case Ω is finite ($\Omega = \{\omega_1, \omega_2, ..., \omega_n\}$), the relation between the basic belief assignment m and its associated belief Bel is expressed algebraically as

$$\text{Bel} = \begin{pmatrix} 1 & 0 \\ 1 & 1 \end{pmatrix} \otimes \begin{pmatrix} 1 & 0 \\ 1 & 1 \end{pmatrix} \otimes ... \otimes \begin{pmatrix} 1 & 0 \\ 1 & 1 \end{pmatrix} \cdot m$$

$$n \text{ times}$$

where Bel and m are identified with vectors having 2^n components. From here, taking into account that $\begin{pmatrix} 1 & 0 \\ 1 & 1 \end{pmatrix}^{-1} = \begin{pmatrix} 1 & 0 \\ -1 & 1 \end{pmatrix}$, the formula

$$m = \begin{pmatrix} 1 & 0 \\ -1 & 1 \end{pmatrix} \otimes \begin{pmatrix} 1 & 0 \\ -1 & 1 \end{pmatrix} \otimes ... \otimes \begin{pmatrix} 1 & 0 \\ -1 & 1 \end{pmatrix} \cdot \text{Bel}$$

$$n \text{ times}$$

is immediate.

Hence, starting from the values of a belief measure, we retrieve the values of the basic belief assignment by "inclusion-exclusion" type formula:

$$m(A) = \sum_{B \subseteq A} (-1)^{|A|-|B|} \text{Bel}(B), \text{ where } |X| = \text{cardinal of } X.$$

Particularly, for p, q and r elementary propositions, we get

$m(p \vee q) = -\text{Bel}(p) - \text{Bel}(q) + \text{Bel}(p \vee q)$,

$m(p \vee q \vee r) = \text{Bel}(p) + \text{Bel}(q) + \text{Bel}(r) - \text{Bel}(p \vee q)$
$\qquad\qquad - \text{Bel}(p \vee r) - \text{Bel}(q \vee r) + \text{Bel}(p \vee q \vee r)$.

For example, let us toss a coin. A head is obtained? Following Dempster-Shafer theory, we obtain the measures

$\text{Bel}(head) = 0$ and $\text{Bel}(\neg head) = 0$

because we have no information about the experiment (the coin may be counterfeit!).

Suppose that a human expert confirms – with the degree of belief 95% – that the coin is "correct" (i.e. the expert is "sure 95%" that $P(head) = 0.5$). Now, Dempster–Shafer theory provides us (as we will see it in the next section) with the following values:

$\text{Bel}(head) = 0.95 \cdot 0.5 = 0.475$ and also $\text{Bel}(\neg head) = 0.475$.

We missed the 5% that the expert cannot certify and that has to be assigned to the "universe" $head \vee \neg head$!

7.4 The Pignistic Transform

Suppose $\Omega = \{\omega_1, \omega_2\}$ and the basic belief assignment possessed by our agent m is confusion-free. Then m is represented by a point M in the triangle ABC in Figure 7.3. When a decision must be made by the agent, he/she will transform the system of beliefs into a system of probabilities $p_1 = P(\omega_1)$, $p_2 = P(\omega_2)$. These probabilities are represented, in Figure 7.3, by a point P on the segment AB.

There is no "absolute" choice of the point P associated to M. However, the pignistic transformation is given as follows:

$$p_1 = m(\{\omega_1\}) + \frac{1}{2} m(\Omega), \quad p_2 = m(\{\omega_2\}) + \frac{1}{2} m(\Omega).$$

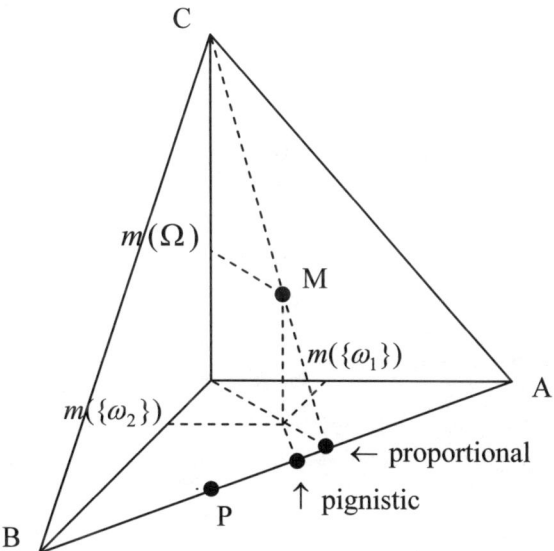

Fig. 7.3. Transforming beliefs into probabilities

In general, the pignistic transform of the b.b.a. $m : 2^\Omega \rightarrow [0,1]$, justified in [Smets 1994], is the probability $P_m : \Omega \rightarrow [0,1]$ given by:

$$P_m(\omega) = \frac{1}{1 - m(\emptyset)} \sum_{\omega \in X \subseteq \Omega} \frac{1}{|X|} m(X),$$

where $|\cdot|$ denotes, as above, the cardinality.

Of course, there are other "rational" transforms, such as the "proportional" one, illustrated in the Figure above.

7.5 Combining Beliefs. The Dempster's Formula

We would like to combine the obtained evidence from two independent sources (for example from two human experts) when the basic belief assignments are expressed.

We start by analyzing how two different basic belief assignments m_1 and m_2 can be combined to obtain a common basic belief assignment m.

We know that $\sum_b m_1(b) = 1$ and $\sum_c m_2(c) = 1$. From here, by multiplication,

$$\sum_{b,c} m_1(b) \cdot m_2(c) = 1 .$$

We regroup the sum's terms in two categories:

1) those which correspond to the subsets B, C such that $B \cap C = \varnothing$;
2) the others, which correspond to the subsets B, C such that $B \cap C = A \neq \varnothing$.

We obtain:

$$\sum_{B \cap C = \varnothing} m_1(B) \cdot m_2(C) + \sum_{A \neq \varnothing} \sum_{B \cap C = A} m_1(B) \cdot m_2(C) = 1 .$$

If we denote by K the first term of this sum, we get:

$$\sum_{A \neq \varnothing} \sum_{B \cap C = A} m_1(B) \cdot m_2(C) = 1 - K ,$$

which means that we denote $m(A) = \dfrac{1}{1-K} \sum_{B \cap C = A} m_1(B) \cdot m_2(C)$; then m satisfies the condition (BBA2) imposed to basic belief assignments. In order to obtain a genuine basic belief assignment, it would suffice to impose the additional condition

$$m(\varnothing) = 0 .$$

This is precisely the formula proposed in [Dempster 1967] for the combination of basic belief assignments.

Hence, starting from two basic belief assignments m_1 and m_2, we compute the (discordance) coefficient

$$K = \sum_{B \cap C = \varnothing} m_1(B) \cdot m_2(C) \left(= \sum_{b \wedge c \Leftrightarrow \text{contradiction}} m_1(b) \cdot m_2(c) \right),$$

then we define the composition $m = m_1 \oplus m_2$ of m_1 and m_2 by

$m(\varnothing) = 0$ and

$$m(A) = \frac{1}{1-K} \sum_{B \cap C = A} m_1(B) \cdot m_2(C) \text{ for } A \neq \varnothing$$

$$\left(\text{res.}\, m(a) = \frac{1}{1-K} \sum_{b \wedge c \Leftrightarrow a} m_1(b) \cdot m_2(c) \text{ for } a \neq \text{contradiction} \right).$$

Therefore, to combine two beliefs Bel_1 and Bel_2, from two independent sources, we consider the corresponding basic belief assignments m_1 and m_2, we compute the composition $m = m_1 \oplus m_2$ and, from here, we compute the composed belief Bel.

Example 1 ([Luger 2002]). Suppose that universe Ω has four atomic propositions, which are the following hypotheses: the patient has a cold (l), flu (f), migraine headache (h), or meningitis (g). We measure the belief of each subset of Ω according to the symptoms.

Obviously, we have to distribute the "unit belief" on $2^4 - 1 = 15$ different subsets.

Suppose the first piece of evidence is clear, our patient has fever. Any hypothesis, except the headache, may be the cause. Our physician, informed by phone, evaluates $m_1(l \vee f \vee g) = 0.6$. If additional information is not available, then the rest of the belief (0.4) has to be distributed to the universe: $m_1(\Omega) = 0.4$.

Suppose that another physician (who is not experienced) obtains some other data for diagnosis, namely that our patient has extreme nausea. Now the causes may be l, f or h. This physician independently evaluates $m_2(l \vee f \vee h) = 0.7$ (and, of course, $m_2(\Omega) = 0.3$).

Knowing both evaluations, we use Dempster's rule to combine these two beliefs. The discordance coefficient is $K = \sum_{B \cap C = \varnothing} m_1(B) \cdot m_2(C) = 0$, so we obtain $m'(X) = \sum_{B \cap C = X} m_1(B) \cdot m_2(C)$ for all $\varnothing \neq X \subseteq \Omega$.

Only for four subsets

$$L \cup F, \; L \cup F \cup H, \; L \cup F \cup G, \; \Omega = L \cup F \cup H \cup G$$

we obtain non-zero values:

$$m'(l \vee f) = m_1(l \vee f \vee g) \cdot m_2(l \vee f \vee h) = 0.42,$$
$$m'(l \vee f \vee g) = m_1(\Omega) \cdot m_2(l \vee f \vee g) = 0.18,$$
$$m'(l \vee f \vee h) = m_1(\Omega) \cdot m_2(l \vee f \vee h) = 0.28 \text{ and}$$
$$m'(\Omega) = m_1(\Omega) \cdot m_2(\Omega) = 0.12.$$

Suppose a new piece of evidence is obtained, this time a fact from the lab; namely, the result of a lab culture is strongly associated with meningitis:

$$m_3(g) = 0.8 \text{ (and } m_3(\Omega) = 0.2 \text{)}.$$

Dempster's formula can help to redistribute the beliefs. Now the discordance coefficient is

$$K = \sum_{B \cap C = \varnothing} m'(B) \cdot m_3(C) = m'(L \cup F) \cdot m_3(G)$$
$$+ m'(L \cup F \cup H) \cdot m_3(G) = 0.42 \cdot 0.8 + 0.28 \cdot 0.8 = 0.56$$

therefore we have to normalize with the coefficient $\dfrac{1}{(1 - 0.56)} = \dfrac{1}{0.44}$ the following results:

$$0.44 \cdot m''(G) = \sum_{B \cap C = G} m'(B) \cdot m_3(C)$$
$$= m'(L \cup F \cup G) \cdot m_3(G) + m'(\Omega) \cdot m_3(G) = ... = 0.24,$$

$$0.44 \cdot m''(L \cup F) = \sum_{B \cap C = L \cup F} m'(B) \cdot m_3(C) = m'(L \cup F) \cdot m_3(\Omega) = 0.084$$

$$...$$

so $m''(g) = 0.545...$, $m''(l \vee f) = 0.190...$, etc.

Notice that the quite large value of discordance coefficient K, 0.56, shows that there are enough conflictual evidence in the choice of basic belief assignments m_1, m_2 and m_3.

Example 2 ([Voorbraak 1993]). Jaundice is a symptom of one of four diseases:

cirrhosis of the liver (l),
hepatitis (h),
gallstones (g),
cancer of gall bladder (k)

This may be wrong, but whenever a set of facts and rules is established, it could be wrong! In Dempster–Shafer theory one assumes that the set of hypotheses is exhaustive; therefore, we should assume a fifth "disease":

any other non-specific liver/gall bladder disorder (n).

Now l and h are liver problems, on the other side g and k are gall bladder problems.

Suppose we have some evidence (for example sharp pain in the lower back) that is associated with liver problem. A first basic belief assignment could be as follows:

$$m_1(l \vee h) = 0.7, \; m_1(l \vee h \vee g \vee k \vee n) = 0.3,$$

and, of course, $m_1(p) = 0$ for all the other propositions p.

Suppose some other evidence is directly available (for example, our patient is not a heavy drinker) that downgrades the hypothesis h. The new basic belief assignment is now the following:

$$m_2(l \vee g \vee k) = 0.8, \; m_2(l \vee h \vee g \vee k \vee n) = 0.2,$$

and $m_2(p) = 0$ for all the other propositions p.

What can be said about the compound mass $m_1 \oplus m_2(l)$?

The only pair b, c such that $m_1(b) \neq 0$, $m_1(c) \neq 0$ and $l \Leftrightarrow b \vee c$ is $b = l \vee h$, $c = l \vee g \vee k \vee n$, so that $\sum_{b \wedge c \Leftrightarrow l} m_1(b) \cdot m_2(c) = 0.7 \cdot 0.8 = 0.56$.

Because the discordance coefficient K is 0, we conclude

$$m_1 \oplus m_2(l) = 0.56$$

despite the fact that we have no idea how the mass 0.7 is distributed among l and h, or how the mass 0.8 is distributed among l, g, and k.

Suppose that "our patient was not a heavy drinker" is interpreted as evidence against the fact that he has liver problems:

$m'_2(g \vee k) = 0.8$, $m'_2(l \vee h \vee g \vee k \vee n) = 0.2$,

$m'_2(p) = 0$ for all the other propositions p.

This time we will get a null value for $m_1 \oplus m'_2(l)$. Instead, the discordance coefficient is $K = m_1(l \vee h) \cdot m'_2(g \vee k) = 0.7 \cdot 0.8 = 0.56$, and the compound masses are:

$$m(l \vee h) = \frac{1}{1-K} m_1(l \vee h) \cdot m'_2(l \vee h \vee g \vee k \vee n) = \frac{0.14}{0.44} \approx 0.32,$$

$$m(g \vee k) = \frac{1}{1-K} m_1(l \vee h \vee g \vee k \vee n) \cdot m'_2(g \vee k) = \frac{0.24}{0.44} \approx 0.54.$$

Notice that these values do not sum up to 1! The difference in mass is, of course, assigned to the universe $l \vee h \vee g \vee k \vee n$.

Example 3 ([Klir and Yuan 1995]). Assume that an old painting was discovered. Two experts estimate the beliefs, starting from the following three atomic propositions:

r = it is a genuine painting by Raphaël,
d = it is a product of one of Raphaël's many disciples,
f = it is a counterfeit.
The beliefs are as follows:

Proposition p	r	d	f	$r \vee d$	$r \vee f$	$d \vee f$
Expert 1 ($\text{Bel}_1(p)$)	0.05	0	0.05	0.2	0.2	0.1
Expert 2 ($\text{Bel}_2(p)$)	0.15	0	0.05	0.2	0.4	0.1

Notice both experts do not "believe" d. By using the inclusion-exclusion formula, we retrieve the mass distributions (of focal propositions):

Proposition p	r	f	$r \vee d$	$r \vee f$	$d \vee f$	$r \vee d \vee f$
$m_1(p)$	0.05	0.05	0.15	0.1	0.05	0.6
$m_2(p)$	0.15	0.05	0.05	0.2	0.05	0.5

Let us compute the composition of basic belief assignments. We start by obtaining the value of discordance coefficient:

$$K = m_1(r) \cdot m_2(f) + m_1(r) \cdot m_2(d \vee f) + m_1(f) \cdot m_2(r)$$
$$+ m_1(f) \cdot m_2(r \vee d) + m_1(r \vee d) \cdot m_2(f) + m_1(d \vee f) \cdot m_2(r) = 0.03 .$$

Thus, $1 - K = 0.97$, i.e. we have strong concordance. Now we can compute the masses:

$$m(r) = \frac{1}{0.97} (m_1(r) \cdot m_2(r) + m_1(r) \cdot m_2(r \vee d) + m_1(r) \cdot m_2(r \vee f) +$$
$$+ m_1(r) \cdot m_2(r \vee d \vee f) + m_1(r \vee d) \cdot m_2(r) + m_1(r \vee d) \cdot m_2(r \vee f)$$
$$+ m_1(r \vee f) \cdot m_2(r) + m_1(r \vee f) \cdot m_2(r \vee d) + m_1(r \vee d \vee f) \cdot m_2(r))$$

...

The results (three digits precision) are presented in the table:

p	r	d	f	$r \vee d$	$r \vee f$	$d \vee f$	$r \vee d \vee f$
$m(p)$	0.214	0.010	0.095	0.116	0.196	0.059	0.309

Because of the context, proposition d = "it is a product of one of Raphaël's many disciples" is now focal!

Now the beliefs and plausibilities of propositions are as follows:

Proposition p	r	d	f	$r \vee d$	$r \vee f$	$d \vee f$
Belief Bel(p)	0.214	0.010	0.095	0.340	**0.505**	0.165
Plausibility Pl(p)	0.835	0.495	0.660	0.915	**0.990**	0.786

Therefore, the maximum belief (and the maximum plausibility) belongs to the proposition "is a Raphaël or is a counterfeit". Perhaps the third expert opinion is necessary!

7.6   Difficulties with Dempster-Shafer's Theory

Dempster–Shafer theory is an example of algebra for the treatment of belief. However, there are situations when the results seem not natural.

A suggestive example was given in [Zadeh 1984]. Zadeh considered the situation of two physicians consulting a patient. Independently, they establish the following basic belief assignments:

Diagnostic d	meningitis	tumor	concussion
Physician 1 ($m_1(a)$)	0.99	0.01	0
Physician 2 ($m_2(a)$)	0	0.01	0.99

They are not in agreement on the main diagnostic. If we use Dempster's rule, then we have $K = 0.9999$, then $m(\text{meningitis}) = m(\text{concussion}) = 0$, et $m(\text{tumor}) = 1$. Therefore, the combined belief in "tumor" – which is a minor diagnostic for both physicians, will be of 100%. This is not natural!

The drawbacks of Dempster–Shafer theory include (a) the fact that to a minor hypothesis one can attach a large belief, and (b) the tedious calculus involved by combination formulas.

7.7 Specializations and the Transferable Belief Model

The previous approach using function(s) Bel is rather static. A new piece of evidence, received and "learned" by the agent(s), could determine a change, even drastic, in their systems of belief. The transferable belief model, as introduced in [Smets 1988], is able to control such changes.

In fact, the transferable belief model postulates that the impact of a new piece of evidence consists in reallocating parts of the initial amount of belief among parts of Ω.

The reallocation – which depends on the capacity of interpreting the evidence by the agent, i.e. on his/hers "intelligence" – can be described by a set of numbers $r(B, A)$, which express the proportion of the initial mass belief of A that is transferred to B once the new piece of evidence is taken into account ("learned") by the agent.

Given $A \subseteq \Omega$, the letter B symbolizes an arbitrary part of Ω, the numbers $r(B, A)$ are non-negative and their sum is 1:

$$\sum_{B \subseteq \Omega} r(B, A) = 1 .$$

Therefore, if m_0 is the (non-normalized) basic belief assignment that describes the belief of our agent at time t_0 (before the new piece of evidence is taken into account), then the b.b.a. m_1 that describes the belief of our agent at a later time t_1 (after "learning" the new piece of evidence) will be obtained as follows:

$$m_1(A) = \sum_{B \subseteq \Omega} r(A, B) m_0(B) \text{ for } A \subseteq \Omega .$$

Of course, the formula above shows how the "vector" m_1 is obtained as the product of the "matrix" r and the "vector" m_0.

It seems reasonable to accept that our agent will reallocate the initial mass $m_0(A)$ only on parts of A^2, i.e. to accept that

$$r(B, A) = 0 \text{ for } B \not\subseteq A .$$

---

[2] However, the history (of mathematics) retains other type of reallocations. For example, in a letter addressed to Richard Dedekind, dated June 20, 1877, Georg Cantor writes: "To this question one should give an affirmative answer, even if for many years I considered as true exactly the contrary".

Reallocations of this type are called **specializations** ([Kruse and Schwecke 1990]).

Important examples of specializations are obtained if we consider the Dempster's "rule of conditioning". More precisely, let us accept that the new piece of evidence "learned" by the agent *You* is the following: "it is impossible that the real world ω_0 be outside $X \subseteq \Omega$". If our agent *You* acts rationally, probably he/she will reallocate the "old" part of belief $m_0(A)$ entirely to the set $A \cap X$. This corresponds to the specialization c_X – called conditioning by X – defined by

$$c_X(B, A) = \begin{cases} 1 & \text{if } B = A \cap X \\ 0 & \text{otherwise.} \end{cases}$$

As another example, the pignistic transform can be viewed as a specialization b, given by:

$$b(B, A) = \frac{\#B}{\#A} \text{ (obviously, for } A, B \neq \varnothing \text{).}$$

Suppose r and r' are specializations. Thus, for $A \subseteq \Omega$,

$$r(B, A) = r'(B, A) = 0 \text{ for any } B \not\subseteq A$$

and

$$\sum_{B \subseteq \Omega} r(B, A) = \sum_{B \subseteq \Omega} r'(B, A) = 1.$$

The last condition is expressed also as follows

$$u \cdot r = u \cdot r' = u$$

where u is the vector $(1, 1, \ldots, 1)$ with all components 1.

The matrix product $r \cdot r'$ gives rise to a reallocation r'', where

$$r''(B, A) = \sum_{C \subseteq \Omega} r(B, C) \cdot r'(C, A).$$

If $B \not\subseteq A$, for any $C \subseteq \Omega$ we have either $B \not\subseteq C$ – in which case $r(B, C) = 0$, or $B \subseteq C$ – in which case $C \not\subseteq A$ thus $r'(C, A) = 0$. Hence $r''(B, A) = 0$.

On the other hand, it is obvious that

$$u \cdot (r \cdot r') = (u \cdot r) \cdot r' = u \cdot r' = u$$

and the following result is proved.

Proposition 7.1. If r and r' are specializations, then $r \cdot r'$ is also a specialization.   ∎

This result is easily interpreted in dynamic terms. Suppose our agent *You* possessed, at initial time t_0, an "evidential corpus" E_0, which lead him to a basic belief assignment m_0. At time t_1, after "learning" new evidence E_1, the agent reallocates beliefs such that the new basic belief assignment is

$$m_1 = r' \cdot m_0 .$$

Later on, at time t_2, after "learning" another new evidence E_2, the new basic belief assignment of our agent becomes

$$m_2 = r \cdot m_1 .$$

Hence $r \cdot r'$ corresponds to the reallocation of beliefs determined by the compound new evidence " E_1 and E_2 ".

Thus, assimilating both pieces of evidence E_1 and E_2 will "expand" the evidential corpus to E . It seems natural to accept that the "expansion" of the evidential corpus does not depend on the order the new pieces of evidence are "learned" (taken into account). However, this leads to a conclusion

$$r' \cdot r = r \cdot r' ,$$

which is not satisfied by all specializations!

A **Dempsterian specialization** s is a specialization that commutes with all the other Dempsterian specializations.

This definition is fully justified if all the possible conditionings are imposed as Dempsterian specializations.

Proposition 7.2. ([Klawoon 1992]). Every Dempsterian specialization s is uniquely determined by a basic belief assignment m such that

$$s(B, A) = \sum_{C \cap A = B} m(C) .$$

Indeed, given s, consider $m(X) = s(X, \Omega)$. The relation between s and m follows from the commutativity conditions $s \cdot c_A = c_A \cdot s$, where c_A is the conditioning on A.   ∎

As a simple example, when the "universe" is $\Omega = \{\omega_1, \omega_2\}$, Dempsterian specializations are described by matrices

$$
\begin{array}{c}
 \\
\varnothing \\
\{\omega_1\} \\
\{\omega_2\} \\
\Omega
\end{array}
\begin{array}{cccc}
\varnothing & \{\omega_1\} & \{\omega_2\} & \Omega \\
\left(1 \right. & \alpha + \gamma & \alpha + \beta & \alpha \\
 & \beta + \delta & & \beta \\
 & & \gamma + \delta & \gamma \\
 & & & \left. \delta \right)
\end{array}
\quad \text{where } \alpha + \beta + \gamma + \delta = 1
$$

which are convex combinations of tensor products

$$\begin{pmatrix} 1 & 1 \\ 0 & 0 \end{pmatrix} \otimes \begin{pmatrix} 1 & 1 \\ 0 & 0 \end{pmatrix}, \begin{pmatrix} 1 & 1 \\ 0 & 0 \end{pmatrix} \otimes \begin{pmatrix} 1 & 0 \\ 0 & 1 \end{pmatrix}, \begin{pmatrix} 1 & 0 \\ 0 & 1 \end{pmatrix} \otimes \begin{pmatrix} 1 & 1 \\ 0 & 0 \end{pmatrix} \text{ and } \begin{pmatrix} 1 & 0 \\ 0 & 1 \end{pmatrix} \otimes \begin{pmatrix} 1 & 0 \\ 0 & 1 \end{pmatrix}.$$

In general, when Ω has n elements, Dempsterian specializations are described by matrices that are convex combinations of 2^n tensor products

$$M_1 \otimes M_2 \otimes \ldots \otimes M_n,$$

where each M_k is either $\begin{pmatrix} 1 & 1 \\ 0 & 0 \end{pmatrix}$, or $\begin{pmatrix} 1 & 0 \\ 0 & 1 \end{pmatrix}$.

7.8 Conditional Beliefs and the Generalized Bayesian Theorem

Beliefs generalize classical probabilities, thus it is natural to extend reasoning in Bayesian networks (see Chapter 5) to beliefs. A definition of conditional beliefs is needed.

Suppose the universe U is a Cartesian product $\Omega \times \Xi$. The Dempster-Shafer approach to treat beliefs of an agent You is difficult, because the subsets of U, in general, are not Cartesian products $A \times X$ with $A \subseteq \Omega$ and $X \subseteq \Xi$.

However, this is the case of a medical diagnosis process. Here Ω is the set of symptoms and Ξ is the set of diseases. It is known that

$$\text{diseases } X \subseteq \Xi \xrightarrow{\text{lead to}} \text{symptoms } A \subseteq \Omega.$$

Suppose for each disease $\xi \in \Xi$ our agent-physician entertains a belief system, over the symptoms Ω, which will be denoted by $\mathrm{bel}_\Omega(\bullet \mid \xi)$. The corresponding basic belief assignment, which can be supposed normalized) is denoted by $m_\Omega(\bullet \mid \xi)$. Is there a natural extension of these belief systems to any subset X of Ξ?

The answer to this question was first given in [Smets 1978] (see also [Smets 1993], [Xu and Smets 1996]). Its starting point is a dual of the Dempster's rule of combination (of normalized basic belief assignments m_1, m_2 – see Section 7.5 above), namely the formula

$$(m_1 \circ m_2)(A) = \sum_{B_1 \cup B_2 = A} m_1(B_1) \cdot m_2(B_2).$$

In general, given $A \subseteq \Omega$ and $X \subseteq \Xi$, consider families $\{B_\xi\}_{\xi \in X}$ of subsets of A such that $\bigcup_{\xi \in X} B_\xi = A$. For $\xi \in X$, the value $m_\Omega(B_\xi \mid \xi)$ is defined. Now, the formula

$$(DC) \qquad m_\Omega(A \mid X) = \sum_{\{B_\xi\}} \prod_{\xi \in X} m_\Omega(B_\xi \mid \xi)$$

defines a (normalized) basic belief assignment $m_\Omega(\bullet \mid X)$, thus a belief measure $bel_\Omega(\bullet \mid X) : 2^\Omega \to [0,1]$.

Of course, this approach supposes that the belief measures $bel_\Omega(\bullet \mid \xi)$, for $\xi \in X$, are "mutually independent". This means that if the agent "knows" $\xi \in X$, then the observation of $A \subseteq \Omega$ does not alter the beliefs of the agent about the other subsets $B \subseteq \Omega$.

Suppose our agent-physician, apart for the conditional belief measures $bel_\Omega(\bullet \mid X)$, possesses also an initial belief system regarding the possible diseases, denoted by $Bel_0 : 2^\Xi \to [0,1]$. How these are transferred "rationally" into a belief measure defined on sets of symptoms? In other words, which belief measure $Bel : 2^\Omega \to [0,1]$ will be "rationally" adopted by the agent? An answer to this question is presented in [Smets 1993]: if $m_0 : 2^\Xi \to [0,1]$ is the basic belief assignment associated to the initial belief measure Bel_0, then

$$Bel(A) = \sum_{X \subseteq \Xi} bel_\Omega(A \mid X) \cdot m_0(X) \text{ for } A \subseteq \Omega$$

defines the belief measure induced over the set of symptoms Ω.

Once a subset A of symptoms is confirmed, which belief measure $bel_\Xi(\bullet \mid A) : 2^\Xi \to [0,1]$ should be rationally adopted by the agent? The following result, presented also in [Smets 1993], is known as the Generalized Bayesian Theorem:

Proposition 7.3. The conditional beliefs $bel_\Omega(\bullet \mid X) : 2^\Omega \to [0,1]$, $X \subseteq \Xi$ and $bel_\Xi(\bullet \mid A) : 2^\Xi \to [0,1]$, $A \subseteq \Omega$ are related to each other by the formula:

$$(GB) \qquad bel_\Xi(X \mid A) = \prod_{\xi \in \Xi - X} bel_\Omega(\Omega - A \mid \xi). \qquad \blacksquare$$

7.9 Solved Exercises

1. Treating flu. The Figure 7.4 below presents a decision tree, with utilities on a scale between -100 and 100. Obviously, the value 100 corresponds to spontaneous healing.

Which is the decision the agent *You* will make?

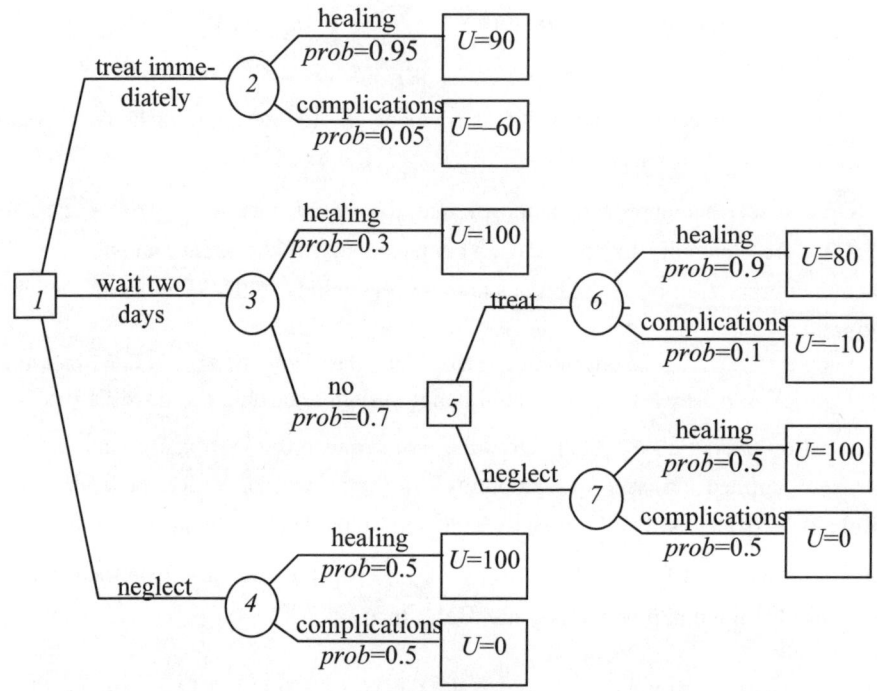

2. Suppose there exists $\lambda > 0$ such that the belief function Bel satisfies the condition (called "the λ-rule"):

$$\text{Bel}(A \cup B) = \text{Bel}(A) + \text{Bel}(B) + \lambda \text{Bel}(A) \cdot \text{Bel}(B)$$

whenever $A, B \subseteq \Omega$, $A \cap B \neq \varnothing$.

Show that, for $E, F \subseteq \Omega$,

$$\text{Bel}(E \cup F) = \frac{\text{Bel}(E) + \text{Bel}(F) - \text{Bel}(E \cap F) + \lambda \text{Bel}(E) \cdot \text{Bel}(F)}{1 + \lambda \text{Bel}(E \cap F)}.$$

3. [Wang and Klir 1992] Suppose m is a normalized basic belief assignment over $\Omega = \{\omega_1, \omega_2, ..., \omega_n\}$ such that $\sum_{k=1}^{n} m(\{\omega_k\}) < 1$. Show that there exists a unique belief function $\text{Bel} : 2^\Omega \to [0, 1]$ that is associated to m and satisfies the λ-rule.

4. Suppose the "old" universe Ω is extended, by taking into account a new possible world $\theta \notin \Omega$, to the "new" universe $\Theta = \Omega \cup \{\theta\}$. Given an "old" belief measure bel associated to the b.b.a. $m : 2^\Omega \to [0, 1]$, two "new" b.b.a.'s $m \downarrow, m \uparrow : 2^\Theta \to [0, 1]$ (hence two "new" belief measures bel $\downarrow$, bel $\uparrow$) can be constructed, as follows:

$$m\!\downarrow\!(X)=\begin{cases}m(X) & \text{if } X\subseteq\Omega\\ 0 & \text{if } \theta\in X\end{cases}, \quad m\!\uparrow\!(X)=\begin{cases}0 & \text{if } X\subseteq\Omega\\ m(X\cap\Omega) & \text{if } \theta\in X\end{cases}.$$

Show that any "new" belief measure $\mathrm{Bel}:2^{\Theta}\to[0,1]$ can be expressed as a weighted average of $\mathrm{bel}_1\!\downarrow$ and $\mathrm{bel}_2\!\uparrow$ for "old" belief measures bel_1, bel_2.

5. Consider a physician (agent) facing a diagnostic problem. The set of possible diseases is $\Xi=\{$bronchitis (b), lung cancer (l), tuberculosis (t)$\}$.

There are two symptoms: the positive X-ray and the fever. The first "universe" is obtained from the ordinary X-ray and has only two elements: $\Omega'=\{$positive X-ray (+), negative X-ray (−)$\}$. The second "universe" has three elements: $\Omega''=\{$no fever at all (no), mild fever (mf), severe fever (sf)$\}$.

Based on his personal experience, the physician is able to form belief systems, conditioned on all the three diseases. These belief systems are represented in tables below by the associated (normalized) b.b.a.'s m', res. m''.

m'	{+}	{−}	{+, −}
•∣b	0.1	0.7	0.2
•∣l	0.4	0	0.6
•∣t	0.8	0	0.2

(Thus a negative X-ray is considered non informative with respect to lung cancer or tuberculosis.)

m''	{no}	{mf}	{no,mf}	{sf}	{no,sf}	{mf,sf}	{no,mf,sf}
•∣b	0	0	0	0.4	0	0.5	0.1
•∣l	0.4	0	0.4	0	0	0	0.2
•∣t	0	0	0	0.1	0	0.6	0.3

Suppose the patient A.B. has severe fever and positive X-ray. Compute the belief system of the physician, and then transform it into probabilities.

6. (The fusion agent.) Suppose each sensor agent Γ_k (from a finite family, of course) possesses, independently from each other, a belief Bel_k about the subsets of the same universe Ω, and conveys this belief to a "fusion agent" Φ. How should this latter agent "fuse" the systems of beliefs?

Solutions
1. Compute iteratively the utilities, starting from leafs toward the root of the decision tree. For a situation like this:

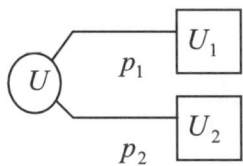

the utility of the "decision of nature" is computed as an average

$$U = p_1 \cdot U_1 + p_2 \cdot U_2 .$$

Hence, $U_6 = 0.9 \cdot 80 + 0.1 \cdot (-10) = 71$, $U_7 = 0.5 \cdot 100 + 0.5 \cdot 0 = 50$,   then $U_4 = 50$ and $U_2 = 0.95 \cdot 90 + 0.05 \cdot (-60) = 81.5$.

Now, for situations like this

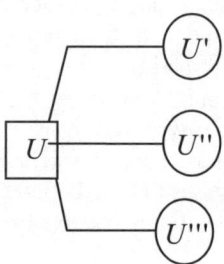

the utility of the "decision" is

$$U = \max\{U',U'',U'''\} .$$

Hence, $U_5 = \max\{71, 50\} = 71$.   Now $U_3 = 0.3 \cdot 100 + 0.7 \cdot 71 = 79.7$,   and finally $U_1 = \max\{81.5, 79.7, 50\} = 81.5$.

The "best" initial decision is that maximizing the utility of U_1, i.e. to treat immediately the flu.

However, the decision to wait two days is the "best" if the probability of spontaneous healing raises from 0.3 to 0.4!

2. Use twice the λ-rule: once for $A = E \cap F$, $B = E - F$ ($A \cup B = E$), and then for $A = E - F$, $B = F$ ($A \cup B = E \cup F$). In particular, for such a belief function,

$$\mathrm{Bel}(\Omega - A) = \frac{1 - \mathrm{Bel}(A)}{1 + \lambda \mathrm{Bel}(A)} .$$

3. Consider the (unique) belief function associated to m. Then for all elements $\omega_k \in \Omega$ one has $\mathrm{Bel}(\{\omega_k\}) = m(\{\omega_k\})$. If Bel satisfies the λ-rule, then by induction $\mathrm{Bel}(A) = \frac{1}{\lambda} \left[\prod_{\omega \in A} (1 + \lambda m(\{\omega\})) - 1 \right]$ and $m(A) \geq 0$. On the other hand, the equation $1 = \mathrm{Bel}(\Omega) = \frac{1}{\lambda} \left[\prod_{\omega \in \Omega} (1 + \lambda m(\{\omega\})) - 1 \right]$ has a unique positive solution if

$$\sum_{k=1}^{n} m(\{\omega_k\}) < 1 .$$

4. Denote $w_1 = \sum_{X \subseteq \Omega} M(X)$ where $M : 2^\Theta \to [0,1]$ is the basic belief assignment associated to Bel.

If $w_1 = 0$, it is clear that $M(X) = 0$ for $X \subseteq \Omega$. Consider, for $A \subseteq \Omega$, $m(A) = M(A \cup \{\theta\})$. Then m is an "old" basic belief assignment, and it is obvious that $M = m \uparrow$, hence $\text{Bel} = \text{bel} \uparrow$.

If $w_1 = 1$, it is clear that $M(X) = 0$ for $\theta \in X$. Obviously $M = m \downarrow$, where $m(A) = M(A)$ for $A \subseteq \Omega$.

If $0 < w_1 < 1$, consider $w_2 = 1 - w_1$, $m_1(A) = \dfrac{1}{w_1} M(A)$ and $m_2(A) = \dfrac{1}{w_2} M(A \cup \{\theta\})$ for $A \subseteq \Omega$. An immediate computation confirms the relation $M = w_1 m_1 \downarrow + w_2 m_2 \uparrow$.

5. The b.b.a.'s presented in the tables are easily converted into conditional belief measures $\text{bel}'(\bullet \,|\, b)$, $\text{bel}'(\bullet \,|\, l)$, $\text{bel}'(\bullet \,|\, t)$, respectively $\text{bel}''(\bullet \,|\, b)$, $\text{bel}''(\bullet \,|\, l)$, $\text{bel}''(\bullet \,|\, t)$. Using (GB), we obtain $\text{bel}_1(\bullet \,|\, +)$ res. $\text{bel}_2(\bullet \,|\, sf)$ as follows:

bel	{b}	{l}	{b, l}	{t}	{b, t}	{l, t}	{b, l, t}
$\text{bel}_1(\bullet \,|\, +)$	0	0	0	0	0	0.7	1
$\text{bel}_2(\bullet \,|\, sf)$	0	0	0	0	0.8	0	1

The associated basic belief assignments are m_1 res. m_2. Using the Dempster's formula, the composition $m = m_1 \oplus m_2$ is

	{b}	{l}	{b, l}	{t}	{b, t}	{l, t}	{b, l, t}
m	0	0	0	0.56	0.24	0.14	0.06

and from here, the associated belief is

	{b}	{l}	{b, l}	{t}	{b, t}	{l, t}	{b, l, t}
Bel	0	0	0	0.56	0.8	0.7	1

Now, the pignistic transform gives as estimates of respective chances, namely $P(t) = 0.77$, $P(b) = 0.14$, $P(l) = 0.09$. The tuberculosis is the first diagnostic at hand.

Notice that replacing the "severe fever" with "mild fever" will produce the (non-normalized) belief

	$\varnothing$	{b}	{l}	{b,l}	{t}	{b,t}	{l,t}	{b,l,t}
$\mathrm{bel}_2(\bullet\mid\mathrm{mf})$	0.016	0.04	0.04	0.1	0.16	0.4	0.4	1

and the final belief

	{b}	{l}	{b, l}	{t}	{b, t}	{l, t}	{b, l, t}
Bel	0.0074	0.0509	0.0695	0.3052	0.3797	0.8139	1

Now, the discordance coefficient is $K = 0.0328$ indicates a weak degree of confusion in the beliefs. The pignistic transform gives the following estimates: $P(t) = 0.601$, $P(b) = 0.080$, $P(l) = 0.319$. Again the tuberculosis is the first diagnostic; however, lung cancer is also probable!

6. (a) The probabilistic approach. A weight γ_k is associated to agent Γ_k such that $\gamma_1 + ... + \gamma_n = 1$. Thus Bel given by

$$\mathrm{Bel}(A) = \gamma_1\mathrm{Bel}_1(A) + ... + \gamma_n\mathrm{Bel}_n(A), \ A \subseteq \Omega$$

is the "weighted" belief measure over Ω. The weights γ_k may express the "importance" given to Γ_k when compared to the other sensor agents.

(b) The belief approach. The fusion agent possesses a belief Bel_0 defined on the subsets of $I = \{1,2,...,n\}$. This belief expresses, for example, the "credibility" of every subset of sensor agents. Each particular belief Bel_k is considered as a conditional belief $\mathrm{bel}(\bullet\mid k)$. The fusion of beliefs is $\mathrm{Bel}(\bullet) = \sum_{K\subseteq I} \mathrm{bel}(\bullet\mid K)\cdot m_0(K)$, where $\mathrm{bel}(\bullet\mid K)$ is computed from $\mathrm{bel}(\bullet\mid k)$ by the rule (DC).

8 Possibility Theory

8.1 Necessity and Possibility Measures

It is not unusual to encounter "possible, but improbable" situations. For example, if X is a standard normal random variable, the event $X = 1$ is possible, but its probability is clearly 0. Let us try an approach of possibilities, as introduced by [Dubois and Prade 1988].

An **uncertainty measure**, defined on a set $\mathcal{P}$ of propositions (containing the "contradiction" false and the "tautology" true, is a function

$$g : \mathcal{P} \to [0,1]$$

satisfying the following three conditions

(U1) $g(\text{false}) = 0$,

(U2) $g(\text{true}) = 1$,

(U3)   If $p, q \in \mathcal{P}$ and q is a logical consequence of p, i.e. if $p \Rightarrow q$, then $g(q) \geq g(p)$.

Since $p \vee q$ is a logical consequence of both p and q, we obtain that g satisfies the inequality

$$g(p \vee q) \geq \max\{g(p), g(q)\}.$$

(Of course, together with p and q, the set $\mathcal{P}$ should contain also $p \vee q$.)

Since both p and q are logical consequences of $p \wedge q$, when $p, q \in \mathcal{P}$ – and supposing that $p \wedge q \in \mathcal{P}$ – we obtain that g satisfies the inequality

$$g(p \wedge q) \leq \min\{g(p), g(q)\}.$$

In general, we have no reason to accept that equality is the correct sign (instead of inequalities) in the last two relations. However, if we impose equality, we obtain so-called possibility res. necessity measures.

A **possibility measure** is a function

$$\Pi : \mathcal{P} \to [0,1]$$

satisfying $\Pi(\text{false}) = 0$, $\Pi(\text{true}) = 1$, $\Pi(p \vee q) = \max\{\Pi(p), \Pi(q)\}$. Of course, as above, we suppose $\text{false} \in \mathcal{P}$, $\text{true} \in \mathcal{P}$ and $p \vee q \in \mathcal{P}$ when $p, q \in \mathcal{P}$.

E. Roventa and T. Spircu: Management of Knowledge Imperfection, STUDFUZZ 227, pp. 187–194.
springerlink.com   © Springer-Verlag Berlin Heidelberg 2009

Notice that the possibility of a union does not depend on the "interaction" of the events expressed by propositions p and q, but only on their possibilities.

Each possibility measure satisfies the following inequality:

$$\Pi(p) + \Pi(\neg p) \geq \max\{\Pi(p), \Pi(\neg p)\} = 1$$

(if p and $\neg p$ both belong to $\mathcal{P}$).

In a dual manner, a **necessity measure** is a function

$$N : \mathcal{P} \to [0, 1]$$

satisfying $N(\text{false}) = 0$, $N(\text{true}) = 1$, $N(p \wedge q) = \min\{N(p), N(q)\}$.

The duality between belief and plausibility in Dempster-Shafer theory can be extended in the general case. Namely, given a possibility measure Π, the formula

$$N(p) = 1 - \Pi(\neg p)$$

defines a necessity measure N. Conversely, once the necessity measure N, is given, the formula

$$\Pi(p) = 1 - N(\neg p)$$

defines a possibility measure Π. This duality allows us to establish that a general necessity measure N satisfies the following inequalities:

$$N(p) + N(\neg p) \leq 1,$$

$$\min\{N(p), N(\neg p)\} = 0.$$

Moreover, several relations involving a pair of dual measures N and Π exist:

a) $N(p) \leq \Pi(p)$,

b) if $\Pi(p) < 1$, then $N(p) = 0$,

c) if $N(p) > 0$, then $\Pi(p) = 1$.

Thus, given dual measures N and Π, the interval $[N(p), \Pi(p)]$ characterizing the uncertainty of a proposition p has always one of its end-points at 0 or at 1.

The total ignorance about p is translated into the conditions $N(p) = 0$, $\Pi(p) = 1$.

When $N(p) = 1$, then also $\Pi(p) = 1$ and p is a certain proposition.

If $N(p) < 1$ and also $N(\neg p) < 1$, then p is an uncertain proposition.

If $N(p) > 0$ and $N(\neg p) > 0$, p is an incoherent proposition.

Let us identify, as in Chapter 7, the propositions in $\mathcal{P}$ to subsets of "universe" Ω. Interpret the negation $\neg$ as the complement, the conjunction $\wedge$ of propositions as the intersection, and the disjunction $\vee$ as the union. The elements $\omega \in \Omega$ correspond to atomic propositions a, i.e. propositions that cannot be expressed as a disjunction $p \vee q$ where both p and q are different from a.

Proposition 8.1. Let Ω be a finite universe and let $\Pi : 2^{\Omega} \rightarrow [0,1]$ be a possibility measure. Then there exists a basic belief assignment $m : 2^{\Omega} \rightarrow [0,1]$ such that Π is exactly the plausibility measure determined by m.

Proof. Suppose $\omega_1, \omega_2, ..., \omega_n$ are all the elements of Ω. Because $\Omega = \bigcup_{k=1}^{n} \{\omega_k\}$ and $\Pi(\Omega) = 1$, we have $1 = \max_k (\Pi(\{\omega_k\}))$. Suppose the elements of Ω are ordered such that $\Pi(\{\omega_1\}) \geq \Pi(\{\omega_2\}) \geq ... \geq \Pi(\{\omega_n\})$. Then $\Pi(\{\omega_1\}) = 1$. Moreover, for $A \subseteq \Omega$ we decompose $A = \bigcup\{\omega_k\}$, hence $\Pi(A) = \max(\Pi(\{\omega_k\})) = \Pi(\{\omega_j\})$, where j is the minimal index such that $\omega_j \in A$.

Consider the following subsets of Ω:

$$C_1 = \{\omega_1\}, \ C_2 = C_1 \cup \{\omega_2\}, \ ..., \ C_n = C_{n-1} \cup \{\omega_n\}.$$

Now, define $m : 2^{\Omega} \rightarrow [0,1]$ as follows:

(a) $m(B) = 0$ for all subsets $B \subseteq \Omega$ other than $C_1, C_2, ..., C_n$;

(b) $m(C_1) = \Pi(\{\omega_1\}) - \Pi(\{\omega_2\})$, ..., $m(C_{n-1}) = \Pi(\{\omega_{n-1}\}) - \Pi(\{\omega_n\})$, $m(C_n) = \Pi(\{\omega_n\})$.

It is obvious that $\sum_{X \subseteq \Omega} m(X) = \Pi(\{\omega_1\}) = 1$.

Denote by Pl the plausibility measure determined by m. The, by definition, $\text{Pl}(A) = \sum_{X \cap A \neq \emptyset} m(X)$ for $A \subseteq \Omega$.

Given $A \subseteq \Omega$, the minimal index j such that $\omega_j \in A$ is exactly the minimal index k such that $C_k \cap A \neq \emptyset$. Thus $\text{Pl}(A) = \sum_{C_k \cap A \neq \emptyset} m(C_k) = \Pi(\{\omega_j\}) = \Pi(A)$, i.e. $\text{Pl} = \Pi$. ∎

The plausibility Pl defined in the frame of Dempster-Shafer theory is an example of a possibility measure:

Proposition 8.2. If $\Omega = \{\omega_1, \omega_2, ..., \omega_n\}$ and the basic belief assignment $m : 2^{\Omega} \rightarrow [0,1]$ is such that $m(X) = 0$ for all $X \subseteq \Omega$ except a "chain" $\emptyset \neq C_1 \subset C_2 \subset ... \subset C_m \subseteq \Omega$, then the plausibility measure Pl determined by m is a possibility measure.

Proof. m is normalized by hypothesis, hence $\text{Pl}(\emptyset) = 0$, $\text{Pl}(\Omega) = 1$. Only the condition $\text{Pl}(A \cup B) = \max\{\text{Pl}(A), \text{Pl}(B)\}$ remains to be proved.

For a given $A \subseteq \Omega$, by hypothesis $\mathrm{Pl}(A) = \sum_{C_k \cap A \neq \varnothing} m(C_k)$. If j_A is minimal among the indices k such that $C_k \cap A \neq \varnothing$, then $\mathrm{Pl}(A) = \sum_{k \geq j_A} m(C_k)$. If j_B is minimal among the indices k such that $C_k \cap B \neq \varnothing$, then $\mathrm{Pl}(B) = \sum_{k \geq j_B} m(C_k)$. Now, if j is minimal among the indices k such that $C_k \cap (A \cup B) \neq \varnothing$, then $j = \min\{j_A, j_B\}$ and $\mathrm{Pl}(A \cup B) = \sum_{k \geq j} m(C_k)$. It is clear now that $\mathrm{Pl}(A \cup B) = \max\{\mathrm{Pl}(A), \mathrm{Pl}(B)\}$.   ∎

The function $\pi : \Omega \to [0,1]$ given by $\pi(\omega) = \Pi(\{\omega\})$ is called the **possibility distribution** associated to the possibility measure $\Pi : 2^\Omega \to [0,1]$. Of course, since $\Pi(A) = \max_{\omega \in A} \pi(\omega)$, Π is determined by π.

For given $\omega_0 \in \Omega$, a possibility distribution $\sigma : \Omega \to [0,1]$ such that

$$\sigma(\omega) = \begin{cases} 1 & \text{for } \omega = \omega_0 \\ 0 & \text{for } \omega \neq \omega_0 \end{cases} \text{ is called "singleton".}$$

If we consider another universe $\Theta = \{\theta_1, \theta_2, ..., \theta_m\}$, then we may take into account possibility measures $T : 2^{\Omega \times \Theta} \to [0,1]$. The associated possibility distribution $\tau : \Omega \times \Theta \to [0,1]$ is in fact a possibilistic relation between the "worlds" in Ω and the "worlds" in Θ. (This is in fact a fuzzy relation – see Chapter 9.)

If π is a possibility distribution defined on Ω, then a possibility distribution defined on Θ appears, given by the formula

$$\rho(\theta) = \max_\omega (\min\{\pi(\omega), \tau(\omega, \theta)\}).$$

This ρ is called the composition of π and τ and is denoted $\rho = \pi \circ \tau$.

Given a possibilistic relation T and an element $\omega_0 \in \Omega$, the **granule** $G(\omega_0) = \sigma_{\omega_0} \circ \tau$ is called the possibility distribution conditioned on ω_0.

8.2 Conditional Possibilities

Let us adopt again the propositional approach. 2^Ω is a Boolean algebra, hence it is natural to accept that the set $\mathcal{P}$ of propositions is a Boolean algebra (with respect to the logical operations $\neg, \wedge, \vee$, and with constants false, true). Consider a possibility measure Π defined on $\mathcal{P}$. The uncertainty of the fact p is expressed by the numbers

$\Pi(p)$ and $\Pi(\neg p)$. At least one of them is 1. The other number expresses the degree of uncertainty. (0 means certain!)

The conditioning in possibility theory is of interest by analogy to the treatment of conditional probabilities.

For a uncertain rule of the form

IF p THEN q

(thus for a uncertain proposition $p \Rightarrow q$) analogous numbers $\pi(q \mid p)$ and $\pi(\neg q \mid p)$ are considered, satisfying the condition

$$\max\{\pi(q \mid p), \pi(\neg q \mid p)\} = 1.$$

These numbers are called **conditional possibilities**.

When the above rule is credible enough, we should have $\pi(q \mid p) = 1$ and $\pi(\neg q \mid p) \geq 0$.

The conditional possibility $\pi(q \mid p)$ may be obtained as the solution of an equation

$$\Pi(p \wedge q) = \min\{x, \Pi(q)\}$$

provided this solution exists and is unique. It is apparent that it is the same formula as in Probability Theory, provided the product is replaced by min.

Of course, the conditional possibility may be defined, in general, as a function $\pi(p \mid q)$ of two arguments p, q (q not "false") satisfying some specific axioms (see [Bouchon-Meunier, Colletti and Marsala 2002]). For example:

(CPoss0) $\pi(q \mid q) = 1$ for every $q \neq$ false;

(CPoss1) $\pi(p \wedge q \mid q) = \pi(p \mid q)$ for $q \neq$ false;

(CPoss2) $\pi(\bullet \mid q) = 1$ is a possibility measure;

(CPoss3) $\pi(p \wedge r \mid q) = \min\{\pi(p \mid q), \pi(r \mid p \wedge q)\}$ for any p, q, r such that $q \neq$ false and $p \wedge q \neq$ false.

Now, for a *Modus ponens* reasoning

IF p THEN q

p

─────────

q

the following matrix formula was suggested:

$$\begin{pmatrix} \Pi(q) \\ \Pi(\neg q) \end{pmatrix} = \begin{pmatrix} \pi(q \mid p) & \pi(q \mid \neg p) \\ \pi(\neg q \mid p) & \pi(\neg q \mid \neg p) \end{pmatrix} \times \begin{pmatrix} \Pi(p) \\ \Pi(\neg p) \end{pmatrix}$$

where the product is replaced by min and the sum by max. Thus for the possibility of the proposition q we should use the formulas

$$\Pi(q) = \max\{\min\{\pi(q \mid p), \Pi(p)\}, \min\{\pi(q \mid \neg p), \Pi(\neg p)\}\}$$

(this will give usually 1) and

$$\Pi(\neg q) = \max\{\min\{\pi(\neg q \mid p), \Pi(p)\}, \min\{\pi(\neg q \mid \neg p), \Pi(\neg p)\}\}.$$

Example. Consider the following premises:

> If John is coming, usually Mary is coming.
> If Mary is coming, sometimes the room is noisy.
> John is coming, almost certainly.

The imprecision is translated into possibility numbers (using the notations m for "Mary is coming", j for "John is coming", n for "room is noisy"):

$$\pi(m \mid j) = 1, \ \pi(\neg m \mid j) = 0.3$$

(0.3 means "usually", see [Simpson 1944], [Negnevitsky 2000]),

$$\pi(m \mid \neg j) = 0 \ (?), \ \pi(\neg m \mid \neg j) = 1$$

(this means that Mary is not coming certainly if John is not coming),

$$\pi(n \mid m) = 0.4, \ \pi(\neg n \mid m) = 1$$

(0.4 means "sometimes"),

$$\pi(n \mid \neg m) = 0 \ (?), \ \pi(\neg n \mid \neg m) = 1$$

(this means that the room is certainly not noisy if Mary is not coming).

Suppose "John is coming, almost certainly". This is translated into:

$$\Pi(t) = 1, \ \Pi(\neg t) = 0.1$$

(0.1 means "almost certainly").

Now we reason twice using the (possibilistic) *Modus ponens* rule

1) *Mary is coming*:
$$\Pi(m) = \max\{\min\{\pi(m \mid j), \Pi(j)\}, \min\{\pi(m \mid \neg j), \Pi(\neg j)\}\}$$
$$= \max\{\min\{1, 1\}, \min\{0, 0.1\}\} = 1,$$
$$\Pi(\neg m) = \max\{\min\{\pi(\neg m \mid j), \Pi(j)\}, \min\{\pi(\neg m \mid \neg j), \Pi(\neg j)\}\}$$
$$= \max\{\min\{0.3, 1\}, \min\{1, 0.1\}\} = 0.3.$$

2) *Room is noisy*:
$$\Pi(n) = \max\{\min\{\pi(n \mid m), \Pi(m)\}, \min\{\pi(n \mid \neg m), \Pi(\neg m)\}\}$$
$$= \max\{\min\{0.4, 1\}, \min\{0, 0.3\}\} = 0.4,$$
$$\Pi(\neg n) = \max\{\min\{\pi(\neg n \mid m), \Pi(m)\}, \min\{\pi(\neg n \mid \neg m), \Pi(\neg m)\}\}$$
$$= \max\{\min\{1, 1\}, \min\{1, 0.3\}\} = 1.$$

From the fact "John is coming, almost certainly", we conclude "The room will be occasionally noisy".

Notice the lack of variation in numbers expressing possibilities: the operations min and max do not create "new" numbers. This rigidity is a major drawback of possibilistic calculus.

Remark. We pointed out above that the uncertainty of the fact p is expressed by two numbers $\Pi(p)$ and $\Pi(\neg p)$. If we want to replace them by a single positive number, call it possit(p) after [Smets 1998], then ½ is a good choice for the complete uncertainty of the proposition p. Suppose that possit(p) $= 1$ will express a (totally) certain proposition, and possit(p) $= 0$ will express that $\neg p$ is (totally) certain. Obviously, the formula

$$\text{possit}(p) = \frac{\Pi(p) + 1 - \Pi(\neg p)}{2} = \frac{\Pi(p) + N(p)}{2}$$

satisfies the requirements above.

The numbers possit are adequate to be compared to the quantification of imprecise terms ([Simpson 1944]). Thus "almost certain" is identified to possit $= 0.95$, "usually" to possit $= 0.85$, "sometimes" to possit $= 0.2$.

8.3 Exercises

1) Draw the possibilistic conclusion from the following propositions:

    (a) When John is ill, rather often one of his colleagues is ill.
    (b) When John is not ill, sometimes one of his colleagues is ill.
    (c) When one of John's colleagues is ill, almost certainly all our colleagues are in danger.
    (d) Today John is, almost certainly, ill.

Notice that "rather often" is identified to possit $= 0.65$.

2) Replace (c), in the list of propositions above, by

    (c') When one of John's colleagues is ill, there are 80% chances for our colleagues to be contaminated.

Draw a possibilistic-probabilistic conclusion.

Solutions

1) Denote j = "John is ill", h = "one of John's colleagues is ill", and o = "all our colleagues are in danger.

The sentences (a)-(d) are translated into the following conditional possibilities:

    (a) $\pi(h \mid j) = 1$, $\pi(\neg h \mid j) = 0.7$ ("rather often"),
    (b) $\pi(h \mid \neg j) = 0.4$, $\pi(\neg h \mid \neg j) = 1$ ("sometimes"),
    (c) $\pi(o \mid h) = 1$, $\pi(\neg o \mid h) = 0.1$ ("almost certainly"),
    (d) $\Pi(j) = 1$, $\Pi(\neg j) = 0.1$.

In order to compute the possit for the conclusion o, we need to evaluate what happens when none of John's colleagues is ill. If we accept that certainly our colleagues are not in danger (?), the final result $\Pi(o) = 1$, $\Pi(\neg o) = 0.7$ will be interpreted perhaps as "more than half of our colleagues are in danger".

2) Denote now c = "(at least one of) our colleagues are contaminated". The sentences express evaluations with respect to the following simple network

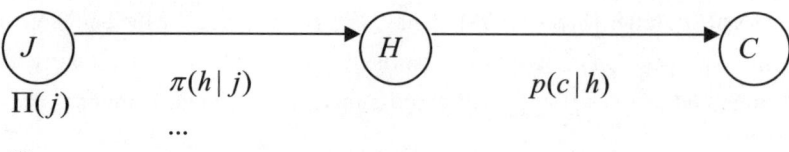

It seems natural to reason like in a Bayesian network, after replacing possibilities with possit values. Thus

$$P(c) = p(c \mid h) \cdot \text{possit}(h) + p(c \mid \neg h) \cdot \text{possit}(\neg h)$$

where, obviously, $\text{possit}(\neg h) = 1 - \text{possit}(h)$.

However, we need an estimate of the probability $p(c \mid \neg h)$. Considering, in a very optimistic evaluation, that $p(c \mid \neg h) = 0$, and knowing $\text{possit}(h) = 0.65$ from Exercise 1, we obtain the estimation $P(c) = 0.8 \cdot 0.65 = 0.52$. There are 52% chances to be contaminated.

9 Approximate Reasoning

9.1 Fuzzy Sets, Fuzzy Numbers, Fuzzy Relations

In classical set theory there is no boundary between a set and its complement; any element x from the universe Ω is either in a given set A, or in its complement $\Omega - A$.

In classical logic (propositional calculus and predicate calculus) only two truth-values of propositions are accepted: true and false.

Neither classical set theory, nor classical logic is characteristic to (ordinary) human thinking. They were used during the last century in order to build and to program computing machines but now, when these machines are powerful enough; it's time to think a little more intelligent.

It is well known now that fuzzy sets were (re)invented by Lotfi Zadeh – see [Zadeh 1965]; since then a lot of people worked in this domain and extended the fuzzy approach to logic.

Let us begin by presenting the main ideas of fuzzy logic. Instead of using only the two truth-values 0 (false) and 1 (true), in fuzzy logic all intermediate (real) values between 0 and 1, i.e. all the values in the segment [0, 1], may be used as genuine truth-values. Thus one accepts that a sentence (proposition) may be partially true.

This situation is rather theoretical. In computer work we cannot use continuum sets, or any other infinite set of truth-values, as for example the dyadic numbers. Thus we have to limit ourselves to a "limited number" of degrees of truth (see Figure 9.1).

Consider the following propositions: "George is tall" and "Vince Carter is tall". (Of course, we mean "height of George is tall" res. "height of Vince Carter is tall".) In classical logic we have to accept these as true or false (depending on a chosen threshold between "tall" and "not tall", let us say 180 cm). Probably each person watching NBA matches will accept the latter as true, but what about the first one?

In fuzzy logic the first proposition may have, for example, the truth-value 0.75. What this means? Perhaps it is accepted that 75% of the people will consider that George is tall, i.e. in a classical way of thinking they will fix the threshold between "tall" and "not tall" below the height of George.

There are many other opinions. Instead of discussing them, it is a better idea to present now the fuzzy sets.

Consider a "universe of discourse" denoted by Ω. In classical set theory, any subset A of Ω is perfectly determined by its characteristic function $\chi_A : \Omega \rightarrow \{0,1\}$, where for any $x \in \Omega$

E. Roventa and T. Spircu: Management of Knowledge Imperfection, STUDFUZZ 227, pp. 195–232.
springerlink.com   © Springer-Verlag Berlin Heidelberg 2009

Classical   Fuzzy (theoretical)   Fuzzy (practical)

Fig. 9.1. Truth values

$$\chi_A(x) = \begin{cases} 1 \text{ if } x \in A \\ 0 \text{ if } x \notin A \end{cases}.$$

In fuzzy set theory, a fuzzy set F is simply a function $\mu_F : \Omega \to [0,1]$. This is called the **membership function** of F because the number $\mu_F(x)$ shows at what extent the element x "belongs" to F.

Of course, if $\mu_F(x) = 0$, then it is accepted that the element x is not in F. All the elements x with $0 < \mu_F(x) < 1$ form a kind of "boundary" of the "core" of F; this **core** is defined as the subset of all x such that $\mu_F(x) = 1$.

A possibility of denoting a fuzzy set (in the universe Ω) is the following:

$$F = \{(x, \mu_F(x)) \mid x \in \Omega\};$$

however, some authors use the reverse notation $F = \{(\mu_F(x) \mid x), x \in \Omega\}$.

Examples. 1) Comfortable house. We may interpret this fuzzy notion in terms of the number of bedrooms in the house. Thus the universe may be the set $\Omega = \{1, 2, 3, ..., 10,...\}$ of natural numbers.

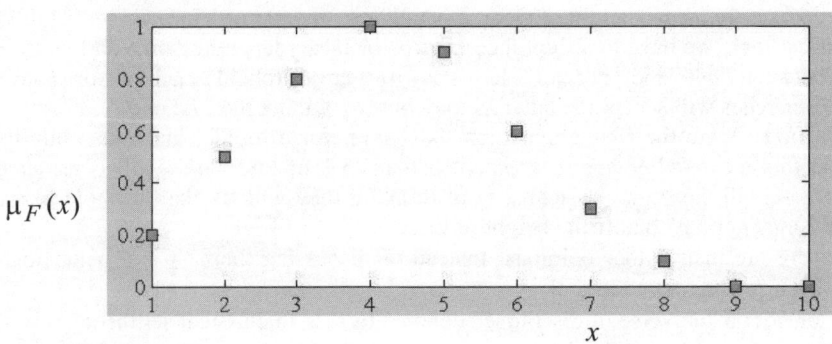

Fig. 9.2. The fuzzy set "comfortable house" for a four-member family

How the fuzzy set F = "comfortable house for a four-member family" may be described? A possibility is presented in the Figure 9.2 above.

2) The expression "around 10" may be considered either as a fuzzy set

$$F = \{(7, 0.1), (8, 0.5), (9, 0.9), (10, 1), (11, 0.9), (12, 0.5), (13, 0.1)\}$$

in the same universe of natural numbers, or as the fuzzy set with membership function:

$$\mu_F(x) = \begin{cases} 0 & \text{if } x \leq 6 \text{ or } x \geq 14 \\ 0.25 \cdot (x - 6) & \text{if } 6 < x < 10 \\ 0.25 \cdot (14 - x) & \text{if } 10 \leq x < 14 \end{cases}$$

in the universe of real numbers. The latter is a triangular fuzzy set.

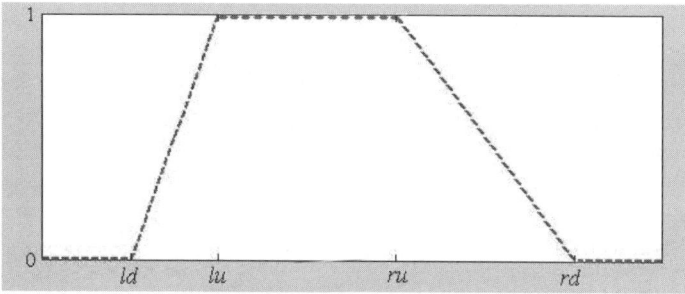

Fig. 9.3. A trapezoidal fuzzy set

A very important family of fuzzy sets, in the universe of real numbers, is formed by the so-called trapezoidal fuzzy sets. A **trapezoidal fuzzy** set (or trapeze) T is determined by four real numbers

$$ld \leq lu \leq ru \leq rd$$

and by the following membership function (see Figure 9.3 above):

$$\mu_T(x) = \begin{cases} 0 & \text{if } x \leq ld \text{ or } x \geq rd \\ \dfrac{x - ld}{lu - ld} & \text{if } ld \leq x \leq lu \\ 1 & \text{if } lu < x < ru \\ \dfrac{rd - x}{rd - ru} & \text{if } ru < x < rd \end{cases}$$

This is implemented as function **trapmf**(x, [ld lu ru rd]) in **Matlab** [1].

---

[1] Software created by The MathWorks, Inc.

Other interesting families of real fuzzy sets, implemented in *Matlab*, are the following:

- Triangular (implemented as function **trimf**),
- S- and Z-shaped (implemented as functions **smf** res. **zmf**),
- Sigmoids (implemented as function **sigmf**, see Figure 9.4a),
- Gaussians (implemented as functions **gaussmf** and **gauss2mf**, see Figure 9.4b),
- Difference of sigmoids (see Figure 9.4c for an example),
- Products of sigmoids and of S-shaped functions.

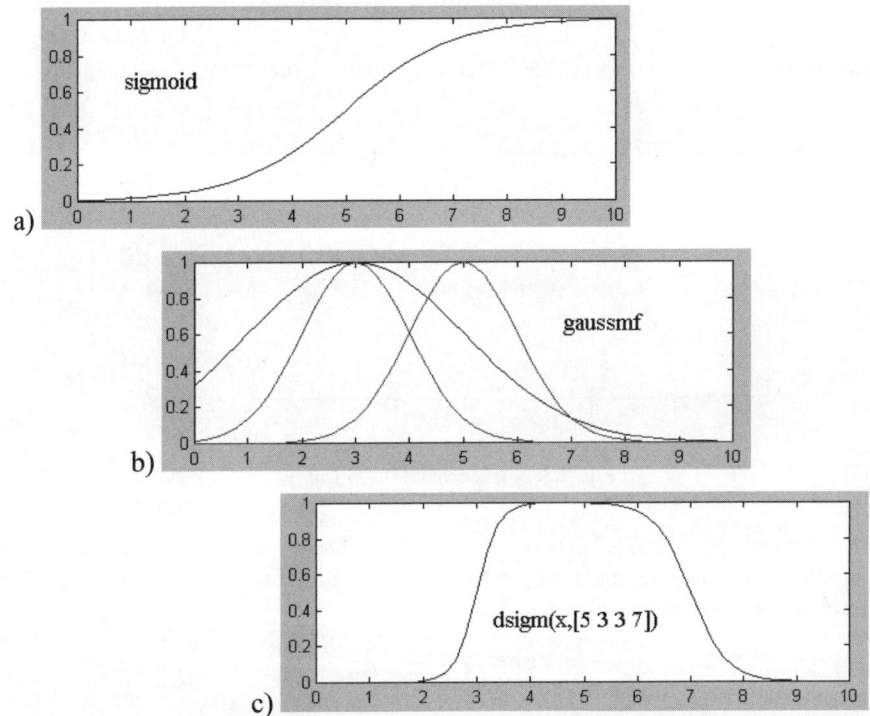

Fig. 9.4. Fuzzy sets implemented in *Matlab*

In general, because $0 \le \mu_F(x) \le 1$ for every fuzzy set F in the universe Ω and for every $x \in \Omega$, it is obvious that the formula

$$\lambda(x) = 1 - \mu_F(x)$$

describes a membership function for a fuzzy set. This fuzzy set is called the **complement** of the fuzzy set F and is denoted $\overline{F}$.

The intersection $F \cap G$ of two fuzzy sets F and G (both in the universe Ω) is defined by the membership function:

$$\mu_{F \cap G}(x) = \min\{\mu_F(x), \mu_G(x)\}.$$

For example, if F = "around 12" and G = "around 11" are described by triangular fuzzy sets, then $F \cap G$ is described as in the Figure 9.5 below. This fuzzy set is not normal (i.e. it has no membership value 1!).

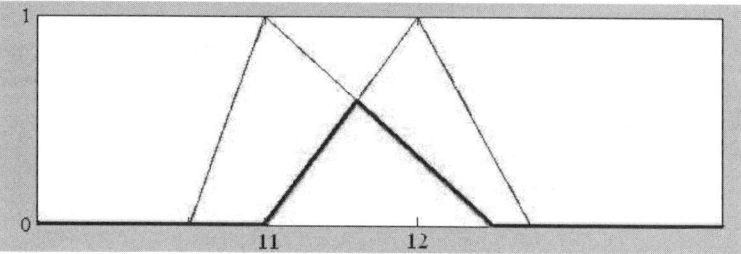

Fig. 9.5. Fuzzy set "intersection"

Dually, the **union** $F \cup G$ is defined by the membership function:

$$\mu_{F \cup G}(x) = \max\{\mu_F(x), \mu_G(x)\}.$$

Other useful operations with fuzzy sets are the following:

- The **algebraic sum** $F + G$, defined by

$$\mu_{F+G}(x) = \mu_F(x) + \mu_G(x) - \mu_F(x) \cdot \mu_G(x).$$

- The **algebraic product** $F \cdot G$, defined by

$$\mu_{F \cdot G}(x) = \mu_F(x) \cdot \mu_G(x).$$

(It is easy to establish that $\overline{F + G} = \overline{F} \cdot \overline{G}$.)

- The **Cartesian product** $F \times H$, defined by

$$\mu_{F \times H}(x, y) = \min\{\mu_F(x), \mu_H(y)\}$$

for $x \in \Omega$, $y \in \Theta$ and H fuzzy set in the universe Θ. (Here $F \times H$ is a fuzzy set in $\Omega \times \Theta$.)

If $\phi : \Omega \to \Theta$ is a function between universes and F is a fuzzy set in Ω, then the formula

$$\mu_H(y) = \begin{cases} \sup\limits_{\phi(x)=y} \{\mu_F(x)\} & \text{if } \phi^{-1}(y) \neq \varnothing \\ 0 & \text{if } \phi^{-1}(y) = \varnothing \end{cases}$$

defines a fuzzy set H in the universe Θ.

Let us consider now fuzzy numbers. By definition, a **real fuzzy number** N is a fuzzy set in the universe of real numbers, satisfying the following conditions:

(FN1)   There is exactly one real number $n \in \mathbf{R}$ such that $\mu_N(n) = 1$.

(FN2) N is min-convex, i.e. for any $x, y \in \mathbf{R}$ and $\lambda \in [0, 1]$

$$\mu_N(\lambda x + (1 - \lambda)y) \geq \min\{\mu_N(x), \mu_N(y)\}.$$

(FN3) μ_N, as a real function, is piece-wise continuous.

What can be said about the operations? For each operation $*$ with real numbers (such as addition, multiplication etc.) a corresponding operation $\circ$ with fuzzy numbers is defined by:

$$\mu_{N \circ M}(x) = \sup_{y * z = x} \{\min\{\mu_N(y), \mu_M(z)\}\}.$$

In particular, the addition $\oplus$ of fuzzy numbers is defined by

$$\mu_{N \oplus M}(x) = \sup_{y + z = x} \{\min\{\mu_N(y), \mu_M(z)\}\}$$

$$= \sup_y \{\min\{\mu_N(y), \mu_M(x - y)\}\}$$

and the multiplication $\otimes$ of fuzzy numbers is defined by

$$\mu_{N \otimes M}(x) = \sup_{y \neq 0} \{\min\{\mu_N(y), \mu_M(x / y)\}\}.$$

The addition $\oplus$ and the multiplication $\otimes$ of fuzzy numbers are commutative and associative. (Although neutral elements could be defined, there is no (natural) opposite, and no natural inverse of a fuzzy number.)

For each function $f : \mathbf{R} \to \mathbf{R}$ and each real fuzzy number N, we define another fuzzy number $f(N)$ by:

$$\mu_{f(N)}(x) = \sup_{x = f(z)} \{\mu_N(z)\}$$

In particular,

$$\mu_{-N}(x) = \sup_{x = -z} \{\mu_N(z)\} = \mu_N(-x)$$

and

$$\mu_{1/N}(x) = \sup_{x = 1/z} \{\mu_N(z)\} = \mu_N(1/x).$$

However – it is important to emphasize – these fuzzy numbers do not possess all the usual properties of the opposite, res. the inverse.

In order to work in a controlled manner with fuzzy numbers, a fourth condition is imposed on N:

$$(\text{FN4}) \qquad \mu_N(x) = \begin{cases} L\left(\dfrac{n-x}{a}\right) & \text{if } x \le n \\[2mm] R\left(\dfrac{x-n}{b}\right) & \text{if } x > n \end{cases}.$$

Here $L(u)$, $R(u)$, the so-called **reference functions**, are supposed to satisfy obvious conditions:

- $L(0) = R(0) = 1$,
- $L(-u) = L(u)$, $R(-u) = R(u)$,
- L, R are decreasing on the interval $[0, +\infty)$.

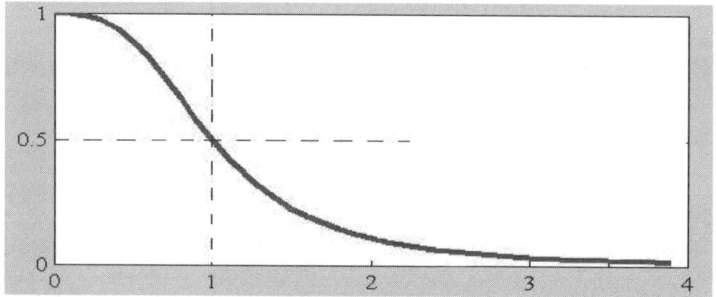

Fig. 9.6. Reference function

As an example,

$$L(u) = R(u) = \frac{1}{1+u^t} \quad \text{where } t = 2, 3, 4.$$

Particular positive values of the constants a, b in (FN4) specify the fuzzy number $N = N(n, a, b)$ "centered" in $n \in \mathbf{R}$. The following formulas are obvious:

$$N(n, a, b) \oplus M(m, c, d) = S(n + m, a + c, b + d),$$

$$N(n, a, b) \otimes M(m, c, d) = P(n \cdot m, n \cdot c + m \cdot a, n \cdot d + m \cdot b)$$
$$\text{for } n, m \ge 0 \ (a, b, c, d > 0).$$

In the "crisp" case a relation R between elements from the set Ω and elements from the set Θ is simply a subset of the Cartesian product $\Omega \times \Theta$. It is possible to compose a relation $R \subseteq \Omega \times \Theta$ with another relation $S \subseteq \Theta \times \Xi$ obtaining a relation $R \circ S \subseteq \Omega \times \Xi$. Namely,

$(x, z) \in R \circ S$ if there exist $y \in \Theta$ such that $(x, y) \in R$ and $(y, z) \in S$.

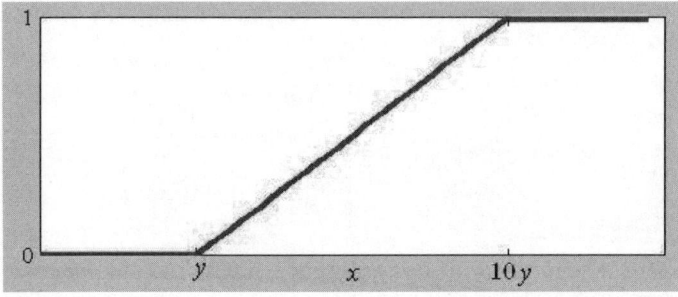

Fig. 9.7. Fuzzy sets "much bigger than"

A **fuzzy relation** R between the (crisp) sets Ω and Θ is a fuzzy set in the universe $\Omega \times \Theta$, i.e.

$$R = \{((x, y), \mu_R(x, y)) \mid x \in \Omega, y \in \Theta\}.$$

Example. Consider $\Omega = \Theta = \mathbf{R}_+$ and the fuzzy relation R = "much bigger than" between real positive numbers, defined by

$$\mu_R(x, y) = \begin{cases} 0 & \text{if } x \le y \\ \dfrac{x - y}{9y} & \text{if } y < x \le 10y \\ 1 & \text{if } x > 10y \end{cases} .$$

This fuzzy relation is represented by real fuzzy sets (depending on y) like in Figure 9.7.

As a particular case, any fuzzy set F in the universe Ω can be assimilated with a relation R between Ω and itself, defined by:

$$\mu_R(x, y) = \begin{cases} 0 & \text{if } y \ne x \\ \mu_F(x) & \text{if } y = x \end{cases} .$$

In the case of finite "small" sets Ω and Θ, a fuzzy relation between the sets Ω and Θ is described as a table of membership function, as for example the following:

μ_R :

Θ / Ω	y_1	y_2	y_3	y_4
x_1	0	0	0.3	0.8
x_2	0.4	0.8	0	1
x_3	0.1	0.7	1	0.9

Fuzzy relations are composed. If $S = \{((y, z), \mu_S(y, z)) \mid y \in \Theta, z \in \Xi\}$ is a fuzzy relation between the sets Θ and Ξ, then $R \circ S$ is the fuzzy relation between Ω and Ξ defined by

$$\mu_{R \circ S}(x, z) = \sup_{y \in \Theta}\{\min\{\mu_R(x, y), \mu_S(y, z)\}\}.$$

Consider as an example the fuzzy relation described by the following table:

μ_S :

Θ ╲ Ξ	z_1	z_2
y_1	0.2	0
y_2	0.9	0.2
y_3	0.7	0.8
y_4	0	0.9

The composition $R \circ S$ is described by the following table (only one value is computed):

$\mu_{R \circ S}$:

Ω ╲ Ξ	z_1	z_2
x_1	.	.
x_2	0.8	.
x_3	.	.

$$\begin{aligned} \max\{&\min\{\mu_R(x_2, y_1), \mu_S(y_1, z_1)\}, \\ &\min\{\mu_R(x_2, y_2), \mu_S(y_2, z_1)\}, \\ &\min\{\mu_R(x_2, y_3), \mu_S(y_3, z_1)\}, \\ &\min\{\mu_R(x_2, y_4), \mu_S(y_4, z_1)\}\} \end{aligned}$$

$$\begin{aligned} &= \max\{\min\{0.4, \ 0.2\}\}, \\ &\quad \min\{0.8, \ 0.9\}, \\ &\quad \min\{0, \ 0.7\}, \\ &\quad \min\{1, \ 0\}\} \\ &= \max\{0.2, \ 0.8, \ 0, \ 0\} = 0.8 \end{aligned}$$

Of course, as a particular case, a fuzzy set F in the universe Ω can be composed with a fuzzy relation R between the sets Ω and Θ. The result is a fuzzy set $F \circ R$ in the universe Θ, defined by:

$$\mu_{F \circ R}(y) = \sup_{x \in \Omega}\{\min\{\mu_F(x), \mu_R(x, y)\}\}.$$

The general definition of fuzzy sets is too coercive. In many applications the membership value $\mu_F(x)$ can be hardly accepted as a "fixed" number, it is rather a fuzzy number. We have to change in some way the definition given above. What was defined as fuzzy set becomes now fuzzy set of type I. A **fuzzy set of type II** is a fuzzy set F, in the universe Ω (of discourse), such that every membership value $\mu_F(x)$ is a fuzzy number.

9.2 Fuzzy Propositions and Fuzzy Logic

Let us return to the problem of assigning truth-values to fuzzy propositions. Consider first the most elementary fuzzy propositions. These are obtained from the generic expressions

$$\mathcal{E}: \text{(linguistic variable) } V \text{ is (fuzzy set) } F$$

In this general expression, F is a fuzzy set associated to the linguistic variable V.

Obvious examples are: temperature is high, wind is strong, speed is slow.

The set of possible values of the linguistic variable V is called the **domain** of V and is denoted $\text{Dom}(V)$. In some fuzzy propositions, values of the variable V are assigned to all individuals from a given set S. Thus the linguistic variable becomes a function

$$V: S \to Dom(V).$$

Given a particular value v of the linguistic variable, suppose we know the membership degree $\mu_F(v)$. This number is interpreted as the truth-value of the particular proposition obtained from $\mathcal{E}$.

The "truth-value" of a generic expression $\mathcal{E}$ is not a number; it depends both on the actual value of the linguistic variable and on the definition (i.e. meaning) of the fuzzy set F.

Consider as an example $V =$ temperature, $F =$ high. Let us represent in Figure 9.8 the membership function identifying the fuzzy set.

When 36°C is the actual temperature, the membership function has value 0.6. Thus the truth-value of the fuzzy proposition

temperature 36°C is high

is 0.6.

The general form of an elementary fuzzy proposition is:

$$P: \text{linguistic value } V(s) \text{ is } F.$$

The truth-value of P is the number $\mu_F(V(s))$; it depends obviously on the individual $s \in S$. This number may be replaced, of course, by a fuzzy number.

Example. $S =$ the set of cars moving toward Toronto on highway 401, $V =$ speed and $F =$ slow. It is clear that the truth-value of proposition

"the speed of a car moving toward Toronto on highway 401 is slow"
is generally "rather false".

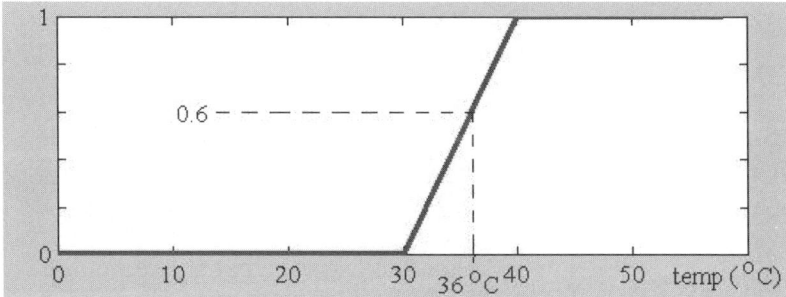

Fig. 9.8. "High temperature"

Human individuals reason using linguistic terms from the natural language. Some of the linguistic terms have a clear meaning; they denote specific objects, locations or actions ("car", "Toronto", "highway 401", "moving toward"). Other terms are ambiguous ("young", "clever", "high", "tall", "slow", "rather false", etc.). The latter fuzzy linguistic terms cannot be used in a strictly bivalent (i.e. based on only two values) reasoning system.

Several fuzzy sets may be associated to a linguistic variable. For example, "young", "adult", "old", "very old" a.o. may be associated to "age". However, for all the fuzzy sets that may be associated to the same linguistic variable, the universe of discourse should be the same. Between the name of the linguistic variable ("age", "intelligence", "temperature", "height", "speed") and the associated fuzzy set in a general expression a semantic relation "is" exist.

As another example, consider the linguistic variable "height" and the following five associated fuzzy sets:

"small", "quite small", "medium", "tall", "very tall"

and the universe of discourse the set $\Omega = [80, 220]$ representing the heights measured in centimeters. Any person defines – in a way or another – these fuzzy sets (an example is in Figure 9.9).

The example above underlines a general feature: terms expressing fuzzy sets are composed of two components: a descriptor of another fuzzy set ("small", "medium", "tall") and a hedge ("", "quite", "very").

From the example "(height of) ... is very tall" it is clear that we can express a generic expression as follows

$$V(\cdot) \text{ is } HD$$

where V is the name of a linguistic variable ("height"), H is a hedge ("very"), D is a descriptor ("tall") of a fuzzy set.

A **fuzzy fact** can be expressed as follows

$$V(o) \text{ is } HD$$

where o is an object (an element from a set). Its truth-value is $\mu_{HD}(V(o))$.

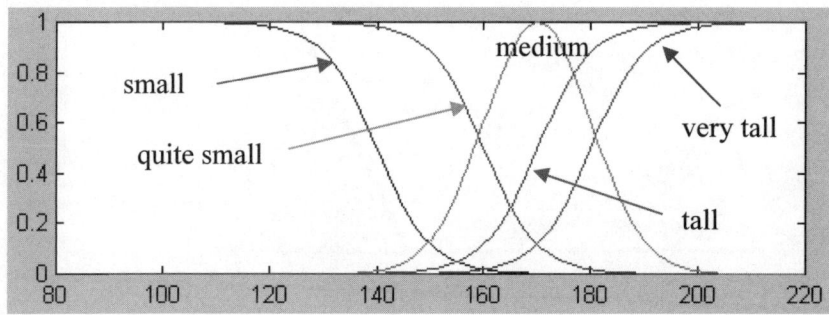

Fig. 9.9. Fuzzy sets associated to "height"

Using logical connectors $\neg$, $\wedge$, $\vee$, $\Rightarrow$ we compose fuzzy propositions from fuzzy facts.

The main idea of fuzzy logic was well expressed by [Zadeh 1973]:

> we want to *approximate* the description of an input/output relation involving two or more variables by a "small" collection of *fuzzy if-then rules*.

Obvious examples of fuzzy if-then rules are:

> z is low if x is high and y is not very low,
> z is low is likely when x is low.

In expressing general fuzzy rules, we may obviously consider the model

> IF *fuzzy antecedent* THEN *fuzzy consequent* (*fuzzy truth-value*).

However, we have to take into account another ingredient, the **fuzzy quantifier**.

In classical logic only two quantifiers are accepted: $\exists$ and $\forall$. By contrast, in fuzzy logic a lot of quantifiers are allowed. Most used are:

"most", "many", "several", "few", "much", "around" etc.

Since the definition by [Zadeh 1965] the "fuzzy" world (sets, numbers, logic) interfered with all the theories created beforehand.

The influence of fuzziness in probabilities is based on the terms *likely* and *probable* and on their derivates: unlikely, highly unlikely, very probable, very likely, not very likely, extremely probable and others.

The influence of fuzziness in possibilities is based on the terms *possible* and *credible* and on their derivates: almost impossible, quite possible, very credible. The influence of fuzziness in beliefs is based on the terms *believable* and *plausible* etc.

9.3 Hedges

Hedges are special linguistic terms by which other linguistic terms (usually expressing fuzzy sets) are modified. Examples: "very", "fairly", "extremely", "slightly", "somewhat", "more or less".

Hedges can be used to modify fuzzy predicates, fuzzy truth-values, fuzzy probabilities, and fuzzy possibilities. For example, in the sentence

"John is possibly young"

we may use up to two copies of the hedge "very", obtaining the following three modified sentences:

"John is possibly very young"
"John is, very possibly, young"
"John is, very possibly, very young"

each one with a different meaning.

In general, given a fuzzy fact

P: $V(o)$ is D

and a hedge H, we can construct a modified fuzzy fact

HP: $V(o)$ is HD

where HD is a new fuzzy set, obtained from fuzzy set D by applying the hedge. What this means? Remember that D, as a fuzzy set, is represented by a membership function

$$\mu_D : \Omega \to [0,1]$$

in the universe of discourse Ω. Consider an increasing one-to-one correspondence $h : [0,1] \to [0,1]$ that represents the hedge. Then the composition

$$h \circ \mu_D : \Omega \to [0,1]$$

is obviously a membership function of a fuzzy set in the same universe of discourse. This fuzzy set is HD.

(Notice D and HD have the same support and the same kernel.)

Three types of hedges exist:

1) Strong, which have the property

$$h(v) < v \text{ for all } v \in (0,1).$$

The most used hedge of this type is "very", defined by the formula $\text{very}(v) = v^2$ (see Figure 9.10).

2) Weak, which have the property

$$h(v) > v \text{ for all } v \in (0,1).$$

The most used hedge of this type is "somewhat", defined by the formula $\text{somewhat}(v) = \sqrt{v}$;

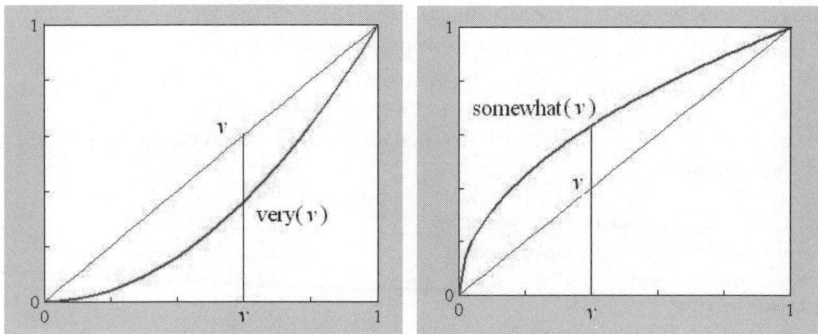

Fig. 9.10. Hedges "very" and "somewhat"

3) Not strong, nor weak. A representative of this type is "indeed", defined by an S-shape formula (see Figure 9.11):

$$\text{indeed}(v) = \begin{cases} 2v^2 & \text{if } v \in [0, \frac{1}{2}] \\ 1 - 2(1-v)^2 & \text{if } v \in (\frac{1}{2}, 1] \end{cases}.$$

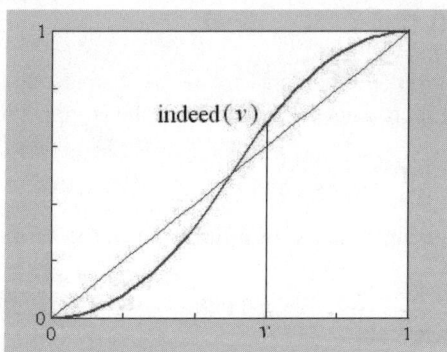

Fig. 9.11. "Indeed" as hedge

Of course, the modified fuzzy fact

HP: $V(o)$ is *HD*

has as truth-value $\mu_{HD}(V(o)) = h(\mu_D(V(o)))$ and, when this is replaced by a fuzzy number, the same (or another) hedge may be applied to this truth-value.

As an example, assume that John is 22 and that the membership function of the fuzzy set "young" is such that $\mu_{young}(22) = 0.9$. Then

"John is young"

has a truth value of 0.9, and

"John is very young"

has a truth-value of 0.81. If the truth-value of the proposition "Jack is possibly young" is evaluated at 0.7, then the truth-value of the proposition "Jack is possibly very young" should be evaluated at 0.49.

The hedges for fuzzy numbers may be defined as follows. First, remember that if n is a real number, then a fuzzy number $N = N(n, a, b)$ above n is represented by the following membership function

$$\mu_N(x) = \begin{cases} L\left(\dfrac{n-x}{a}\right) & \text{if } x \leq n \\ R\left(\dfrac{x-n}{b}\right) & \text{if } x > n \end{cases}$$

where (let us say) $L(u) = R(u) = \dfrac{1}{1+u^2}$.

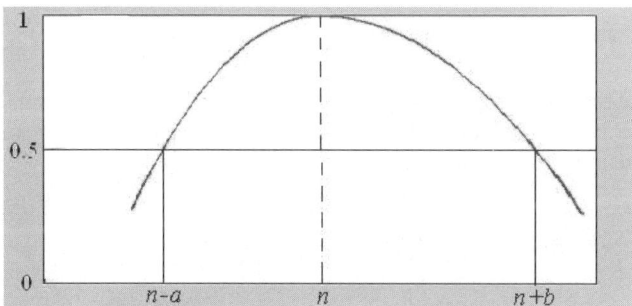

Fig. 9.12. The fuzzy real number $N(n, a, b)$

$L(u) = R(u) = \dfrac{1}{2}$ means exactly $u = 1$. Therefore, we find $x = n - a$ res.

$x = n + b$ as the numbers x where $\mu_N(x) = \dfrac{1}{2}$,

Now, the hedge "nearly" is defined as follows:

nearly$(n) = N(n, 0.05n, 0.05n)$.

Analogously,   about(n) is $N(n, 0.10n, 0.10n)$,

roughly(n) is $N(n, 0.25n, 0.25n)$,

crudely(n) is $N(n, 0.50n, 0.50n)$.

The definition of fuzzy numbers (and the hedges above) allows us to evaluate the membership value in comparisons like this

N is approximately equal *M,*
nearly *n* is approximately roughly *m*

etc. thus to evaluate the truth-values of propositions of this kind. It is easy to notice that the curves representing the compared fuzzy numbers cross at one or two points. The truth-value of a proposition

$$N(n, a, b) \text{ is approximately equal } M(m, c, d)$$

is given by the maximal level of the intersection points and can be calculated from definition formulas.

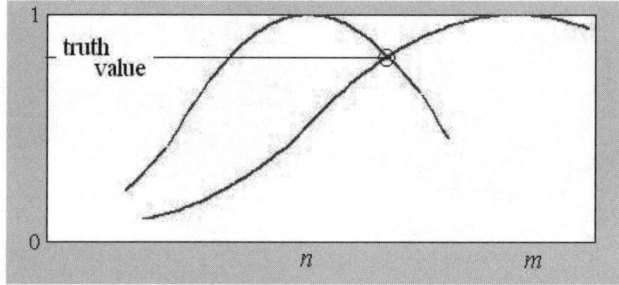

Fig. 9.13. Truth-value of "*N* is approximately equal *M*"

Up to now several models for representing modifiers – for fuzzy sets and fuzzy numbers – have been considered. It is unanimously accepted that, in representing truth-values by using several modifiers before the word "true", such as

(undecided)	$\frac{8}{16} = 0.5$	somewhat	$\frac{9}{16}$	moderately	$\frac{10}{16}$
pretty	$\frac{11}{16}$	really	$\frac{12}{16}$	strongly	$\frac{13}{16}$
very	$\frac{14}{16}$	extremely	$\frac{15}{16}$	(absolutely)	$\frac{16}{16} = 1$

the "numeric" values are not equally-spaced between $\frac{1}{2}$ and 1 (as it was suggested above).

A Bézier transformation is able to convert our modifiers into values in the interval $[\frac{1}{2}, 1]$ which tallies better to our intuition; namely, the values tend to be closer to one another as we move towards "absolutely true" – see Figure 9.15.

A Bézier curve – in a real vector space – is determined by four vectors A, B (called "end nodes") and C, D (called "control nodes") – see Figure 9.14. Each "point" P on such a curve is obtained as a value of the cubic polynomial

$$P = P(\tau) = (1-\tau)^3 A + 3\tau(1-\tau)^2 C + 3\tau^2(1-\tau)D + \tau^3 B$$

where τ is ranging from 0 to 1.

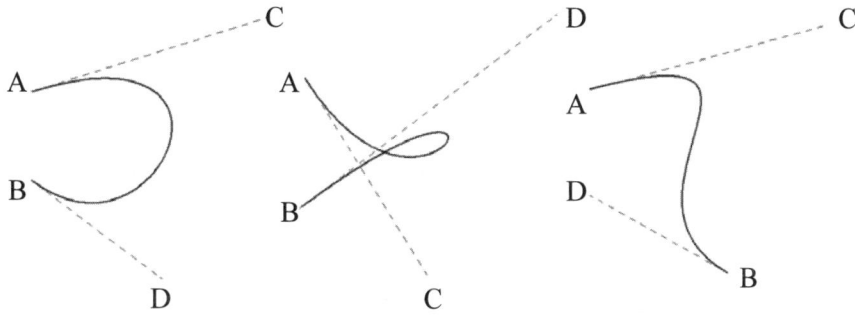

Fig. 9.14. Bézier curves

Consider mainly Bézier curves that are contained in the unit square $[0,1]\times[0,1]$, usually with end nodes $A = (0,0)$ and $B = (1,1)$. For the control nodes $C = (1,0)$ and $D = (\lambda,1)$ we obtain the family called $B(\lambda)$.

As an example, the Bézier curve with end nodes $A = (0,0)$, $B = (1,1)$ and control nodes $C = (\frac{1}{2},0)$, $D = (\frac{1}{2},1)$ can be used to numerically represent intermediate truth values as follows (see Figure 9.15):

somewhat true	0.62	strongly true	0.94
moderately true	0.73	very true	0.97
pretty true	0.82	extremely true	0.99
really true	0.89	(absolutely) true	1

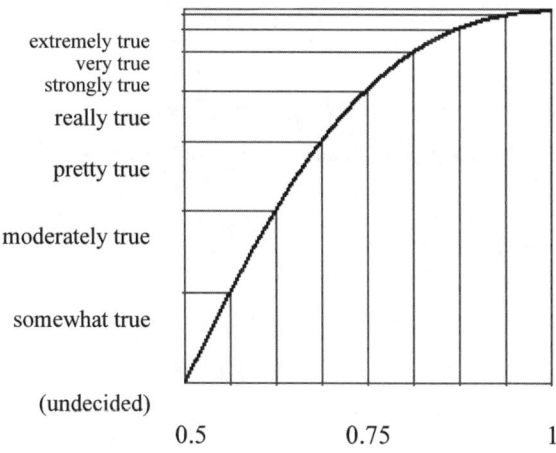

Fig. 9.15. Representing intermediate truth-values

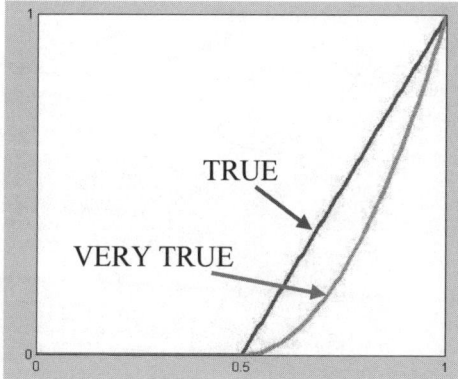

Fig. 9.16. TRUE and VERY TRUE as fuzzy truth-values

Of course, numeric truth-values may be replaced by fuzzy truth-values. For example, "true" as a fuzzy truth-value is well represented by the membership function in Figure 9.16.

In this case "very true" is a fuzzy truth-value obviously different from the numeric value 0.97 above. Possible confusions may appear, due to the imprecision of natural languages. To avoid confusions, it is a good idea to denote fuzzy sets by capital letters.

Quantifiers may be also affected by hedges. Let us list several such hedges: "essentially", "virtually", "practically", "occasionally", "frequently", "largely", "basically", "roughly", "most". Of course, computer programs need the interpretation of such hedges in terms of (either crisp or fuzzy) numbers.

9.4 Fuzzy Logic

The source of approximate reasoning lays in the third truth-value ½ introduced by Łukasiewicz[2], interpreted as "neutral" or "yet undetermined".

The negation, conjunction, disjunction and implication are defined in the ternary logic of Łukasiewicz as follows (here p, q are 0, $\frac{1}{2}$ or 1):

The "compact" formulas that replace the tables below are as follows:

(L1) $\neg p = 1 - p$,

(L2) $p \wedge q = \min\{p, q\}$,

(L3) $p \vee q = \max\{p, q\}$,

(L4) $p \Rightarrow q = \min\{1, 1 - p + q\}$.

We should notice here that the well-known relation in classical logic

$$p \Rightarrow q \text{ is equivalent to } (\neg p) \vee q$$

is no longer correct: the truth-value $\frac{1}{2} \Rightarrow \frac{1}{2}$ is 1, not $\frac{1}{2}$!

---

[2] Jan Łukasiewicz (1878-1956), Polish mathematician.

p	$\neg p$
0	1
½	½
1	0

$p \wedge q$		q	
	0	½	1
p 0	0	0	0
½	0	½	½
1	0	½	1

$p \vee q$		q			$p \Rightarrow q$		q	
	0	½	1			0	½	1
p 0	0	½	1		p 0	1	1	1
½	½	½	1		½	½	1	1
1	1	1	1		1	0	½	1

In Łukasiewicz' ternary logic we drop a lot of other well-known "laws" from classical logic, such as the law of contradiction

$p \wedge \neg p$ is always false

(here $½ \wedge \neg ½$ is $½$, not 0) and the law of excluded middle

$p \vee \neg p$ is always true

(here $½ \vee \neg ½$ is $½$, not 1).

However, some other relation from classical logic, such as

$\neg(p \wedge q)$ is equivalent to $(\neg p) \vee (\neg q)$,

$(p \Rightarrow q) \Rightarrow q$ is equivalent to $p \vee q$,

$p \Rightarrow (q \Rightarrow r)$ is equivalent to $q \Rightarrow (p \Rightarrow r)$

remain true in ternary logic of Łukasiewicz.

Many other multi-valued logics with different properties exist. The following two, satisfying conditions (L1-L3), are known as:

a) the ternary logic of Kleene, in which the implication is defined by

$$(K4) \quad p \Rightarrow_K q = \max\{1 - p, q\}$$

instead of (L4);

b) the ternary logic of Heyting, in which the implication is defined by

$$(H4) \quad p \Rightarrow_H q = \sup_{x \in [0,1]} \{x \mid \min\{p, x\} \le q\}$$

instead of (L4).

The **theoretical fuzzy logic** is a generalization of the ternary logic of Łukasiewicz. The truth-values cover (theoretically!) the interval [0, 1], and the negation, conjunction, disjunction and implication of truth-values p and q are defined by formulas similar to (L1-L4) above. That's why fuzzy logic is a kind of multi-valued logic.

Remember in fuzzy logic we try to emulate patterns of real-life human reasoning. Why exactly the formulas (L1-L4) above should be chosen? In fact, there is no apparent reason; it is possible to replace them by other formulas, as in the cases of Kleene and Heyting logics.

For example, instead of the formula (L1) for the negation we may choose any function $c : [0,1] \rightarrow [0,1]$ satisfying natural conditions:

(N0) $c(0) = 1$ and $c(1) = 0$

(compatibility with classical logic)

(N1)   if $p < p'$ are numbers in [0, 1], then $c(p) > c(p')$

(decreasing condition)

(N2) $c(c(p)) = p$

(involutivity)

or

(N2') $c(1 - p) = 1 - c(p)$

(commuting with classical negation).

A function $c : [0,1] \rightarrow [0,1]$ satisfying (N0), (N1), (N2) or (N2') is called a **fuzzy complement**.

Special examples of fuzzy complements are the Sugeno functions c_λ where

$$c_\lambda(p) = \frac{1-p}{1+\lambda p}$$ for $\lambda \in (-1, +\infty)$, which are involutive, and the function γ given

by $\gamma(p) = \frac{1}{2}(1 + \cos \pi p)$, which commutes with the classical negation.

Of course, suitable Bézier curves may serve as fuzzy complements. The simplest example is the Bézier curve with end nodes $A = (0,1)$, $B = (1,0)$ and the control nodes $C = (1,1)$, $D = (0,0)$. It gives a fuzzy complement that commutes with the classical negation. An involutive fuzzy complement is obtained by replacing the control node D with $(1,1)$.

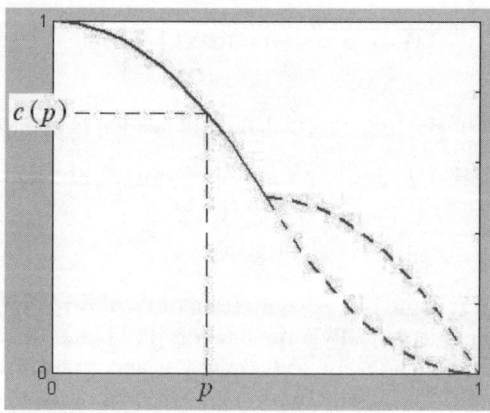

Fig. 9.17. Fuzzy complements

It is immediate that $c(\frac{1}{2}) = \frac{1}{2}$ for any fuzzy complement that commutes with the classical negation. Thus condition (N2') should be considered if compatibility with ternary logic of Łukasiewicz is desired.

Let us consider now the formula $\min\{p, q\}$ that is used in (L2) to define the conjunction $p \wedge q$ when $p, q \in \{0, \frac{1}{2}, 1\}$.

This formula is not the only possibility when trying to define a conjunction $p \wedge q$ for $p, q \in [0, 1]$.

A **fuzzy conjunction** is a function $T : [0, 1] \times [0, 1] \to [0, 1]$ which satisfies the following five conditions:

(T0) $T(0, 0) = T(0, 1) = T(1, 0) = 0$, $T(1, 1) = 1$
  (compatibility with classical logic)

(T1) $T(p, 1) = p$
  (a boundary condition)

(T2) if p and $q < q'$ are numbers in [0, 1], then $T(p, q) \leq T(p, q')$
  (i.e. T is monotonic in the second argument)

(T3) $T(q, p) \leq T(p, q)$
  (i.e. T is commutative)

(T4) $T(p, T(q, r)) = T(T(p, q), r)$
  (i.e. T is associative).

These conditions are adequate to define a genuine "conjunction" by

$$p \wedge q = T(p, q) \text{ for } p, q \in [0, 1].$$

Notice that compatibility with ternary logic of Łukasiewicz exists only if $T(\frac{1}{2}, \frac{1}{2}) = \frac{1}{2}$. This is not a consequence of conditions (T0-T4).

In the literature, functions T satisfying conditions (T0-T4) are known as **t-norms**. Hence, a fuzzy conjunction is simply a t-norm.

Apart from $T(p, q) = \min\{p, q\}$, let us give other three examples:

$$T(p, q) = p \cdot q \text{ (the \textbf{algebraic product}),}$$

$$T(p, q) = \max\{0, p + q - 1\} \text{ (the \textbf{bounded difference}),}$$

$$T(p, q) = \log\left(1 + \frac{(e^p - 1)(e^q - 1)}{e - 1}\right).$$

In a dual manner, let us consider the formula $\max\{p, q\}$ that is used in (L3) to define the disjunction $p \vee q$ when $p, q \in \{0, \frac{1}{2}, 1\}$.

A **fuzzy disjunction** is a function $S : [0, 1] \times [0, 1] \to [0, 1]$ which satisfies the following five conditions, dual to (T0-T4):

(T'0) $S(0, 0) = 0$, $S(0, 1) = S(1, 0) = S(1, 1) = 1$

(T'1) $S(p, 0) = p$

(T'2)   if p and $q < q'$ are numbers in [0, 1], then $S(p,q) \leq S(p,q')$

(T'3) $S(q, p) \leq S(p,q)$

(T'4) $S(p, S(q,r)) = S(S(p,q), r)$.

These conditions are adequate to define a genuine "disjunction" by

$$p \vee q = S(p,q) \text{ for } p, q \in [0,1].$$

Notice conditions (T'2-T'4) are exactly conditions (T2-T4).

In the literature, functions S satisfying conditions (T'0-T'4) are known as **t-conorms**. Hence, a fuzzy disjunction is simply a t-conorm.

It is easy to verify that fuzzy conjunctions and fuzzy disjunctions are deduced one from another by using the formula

$$S(p,q) = c(T(c(p), c(q))) , \ T(p,q) = c(S(c(p), c(q)))$$

where c is an involutive fuzzy complement.

The logic operation of implication is essential for reasoning, either in the classical logic or in fuzzy logic. To define the implication $p \Rightarrow q$ when $p, q \in \{0, \frac{1}{2}, 1\}$, in the logic of Łukasiewicz the formula $\min\{1, 1 - p + q\}$ was used. Other possibilities are given by the formulas $\max\{1 - p, q\}$ and $\max_{x \in [0,1]} \{x \mid \min\{p, x\} \leq q\}$.

These particular implications satisfy different conditions.

A **fuzzy implication** is a function $I : [0,1] \times [0,1] \rightarrow [0,1]$ which satisfies conditions that are adequate to define an implication by

$$p \Rightarrow q = I(p,q) \text{ for } p, q \in [0,1].$$

In classical logic, where $p, q \in \{0,1\}$, $p \Rightarrow q$ is equivalent to $(\neg p) \vee q$. After choosing a fuzzy complement c and a fuzzy disjunction S, we may define the "derived fuzzy implication" as follows:

$$I(p,q) = S(c(p), q) \text{ for } p, q \in [0,1].$$

However, in general this formula does not preserve the above equivalence for $p, q \in \{0, \frac{1}{2}, 1\}$, nor the compatibility with ternary logic of Łukasiewicz is guaranteed.

A fuzzy implication may satisfy some of the following conditions:

(I1) $I(0, p) = 1$

      (dominance of falsity)

(I2) $I(1, q) = q$

      (left neutrality of truth)

(I2') $I(1, q) \geq q$ for $q \geq \frac{1}{2}$, res. $I(1, q) \leq q$ for $q < \frac{1}{2}$

      (see [Dubois and Prade 1991], [Trillas and Valverde 1981])

(I3) $I(p, p) = 1$

      (identity)

(I4)   if $p \leq q$, then $I(p,q) = 1$
       (boundary condition)

(I4') if p and q are numbers in [0, 1], such that $I(p,q) = 1$, then $p \leq q$
       (strict boundary condition)

(I5) $I(q,p) = I(c(p), c(q))$ for a fuzzy complement c
       (the contraposition property)

(I6)   if $p < p'$ and q are numbers in [0, 1], then $I(p,q) \geq I(p',q)$
       (i.e. I is decreasing in the first argument)

(I7)   if p and $q < q'$ are numbers in [0, 1], then $I(p,q) \leq I(p,q')$
       (i.e. I is increasing in the second argument)

(I8) $I(p, I(q,r)) = I(q, I(p,r))$
       (the exchange property)

(I9) $I(p,r) = T(I(p,q), I(q,r))$ for a fuzzy conjunction T.

None of the above conditions is mandatory. For example, condition (I2'), which is an alternative to (I2), is justified by the following observation: when a physician is confronted with a sentence of the form "(absolutely) true $\Rightarrow$ really true", he/she is inclined to consider it as being "very true" rather than the "really true" imposed by (I2).

The exchange property (I8) is unacceptable if we analyze the way a physician is reasoning. In his mind the temporal order is extremely important, "first fact P and then fact Q" usually triggers a completely different action as "first fact Q and then fact P".

At first sight condition (I9) seems natural, imposed by the hypothetical syllogism inference rule. However, it is not satisfied by any of the known fuzzy implications (see [Klir and Yuan 1995])!

Probably most specialists will accept only the conditions (I1), (I5), (I6) and (I7) as genuine for a fuzzy implication. But this is a matter of personal choice.

Examples. From the bounded difference t-conorm (and the classical negation) we derive the fuzzy implication $I(p,q) = \min\{1, 1 - p + q\}$ which extends the Łukasiewicz implication. This satisfies all conditions above, except (I9). From the t-conorm $\max\{p,q\}$ associated to the t-norm $\min\{p,q\}$ we derive the fuzzy implication $I(p,q) = \max\{1 - p, q\}$ which extends the Kleene implication. This implication does not satisfy (I3) and (I4). Also the fuzzy implication $I(p,q) = 1 - p + p \cdot q$ derived from the algebraic product t-norm does not satisfy conditions (I3) and (I4).

It is easy to propose a Bézier-surface function $I(p,q)$ – see Figure 9.18 – satisfying conditions (I1), (I5)-(I7) above. Namely, for $p \in [0,1]$ we first use the end nodes $A_0 = (0,0,1)$, $B_0 = (1,0,0)$ and the control nodes $C_0 = (1,0,1)$, $D_0 = (0,0,0)$ in order to obtain $I(p,0)$, then for $q \in [0, p]$ use the end nodes $A = (p, 0, I(p,0))$, $B = (p, p, 1)$ and the control nodes $C = (p, p, I(p,0))$, $D = (p, 0, 1)$ in order to obtain

$$J(p,q) = \frac{1}{4}(1-\tau)^2(2+\tau)I(p,0) + \frac{1}{4}(1+\tau)^2(2-\tau)$$

where $\tau = \sqrt[3]{2q/p - 1}$ and $I(p,0) = \frac{1}{4}(1-\sigma)^2(2+\sigma)$ with $\sigma = \sqrt[3]{2p-1}$. Finally
$I(p,q) = (J(p,q) + J(1-q,1-p))/2$.

This construction satisfies also conditions (I2'), (I3), (I4), (I4'), smoothness conditions, and also algebraic relations like the following

$$I(p,q) + I(p,p-q) = 1 + I(p,0) \text{ for } p > q.$$

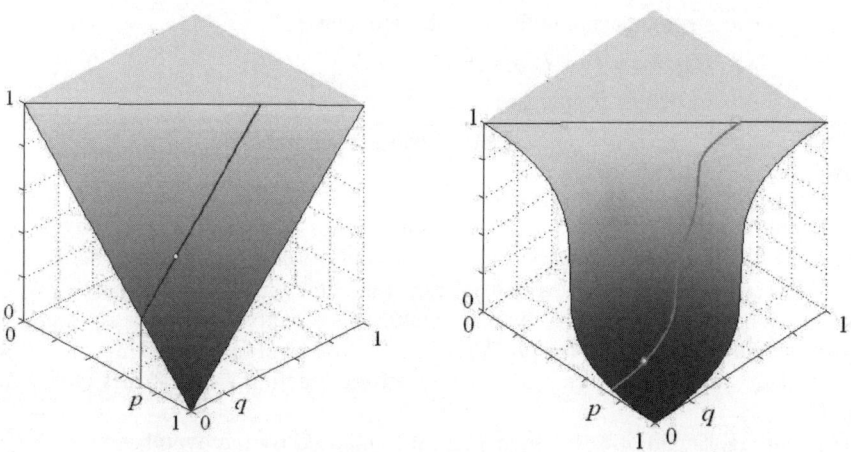

Fig. 9.18. Fuzzy implications: generated by bounded difference (left), generated by Bézier curves (right)

9.5 Approximate Reasoning

Given a t-norm T, T-granules are immediate generalizations of Cartesian products of fuzzy sets. More precisely,

Given a fuzzy set F in the universe Ω and a fuzzy set G in the universe Θ, the **T-granule** determined by the pair (F, G) is the fuzzy set defined by the membership function

$$\mu(x,y) = T(\mu_F(x), \mu_G(y)) \text{ for } x \in \Omega, y \in \Theta.$$

This fuzzy set, in the universe $\Omega \times \Theta$, will be denoted $F \times_T G$.

Let us choose – and keep fixed – an involutive fuzzy complement c, a fuzzy conjunction (t-norm) T, and a fuzzy implication I. Denote by S the fuzzy disjunction (t-conorm) induce by c and T. These will help us to define truth-values of fuzzy propositions composed of fuzzy facts.

More precisely, recall a fuzzy fact P can be expressed as follows

$$P: V(o) \text{ is } F$$

where o is an object, V is a linguistic variable, and F is a fuzzy set in a universe Ω that contains all the values of V. The truth-value of P is

$$\mu_F(V(o)) \in [0,1]$$

thus depends on the object o.

The notation $\neg P$ will denote the fuzzy proposition "not $V(o)$ is F", whose truth-value will be defined as $c(\mu_F(V(o)))$. It is important to make the distinction between the fuzzy proposition $\neg P$ and the following expression "$V(o)$ is not F", which is not a fuzzy fact!

Now, given another fuzzy fact

$$Q: W(u) \text{ is } G$$

where u is an object, W is a linguistic variable, and G is a fuzzy set in a universe Θ that contains the values of W, we may consider the fuzzy propositions

$$V(o) \text{ is } F \text{ and } W(u) \text{ is } G, \text{ denoted by } P \wedge Q,$$

$$V(o) \text{ is } F \text{ or } W(u) \text{ is } G, \text{ denoted by } P \vee Q,$$

whose truth-values will be defined as

$$T(\mu_F(V(o)), \mu_G(V(u))), \text{ res. } S(\mu_F(V(o)), \mu_G(V(u))).$$

The logical operations $\neg, \wedge, \vee$ are easily extended to arbitrary fuzzy propositions and the corresponding truth-values are obtained accordingly.

The set of fuzzy propositions is the minimal set of results of possible constructions made respecting the following:

(FP0)   Fuzzy facts are fuzzy propositions.

(FP1)   If P is a fuzzy proposition, then $\neg P$ is also a fuzzy proposition.

(FP2)   If P and Q are fuzzy propositions, then $P \wedge Q$ and $P \vee Q$ are also fuzzy propositions.

Of course, any fuzzy proposition has a truth-value attached.

Due to the conditions satisfied by c, T and S, some of the well-known laws in classical logic are preserved, some others no. The law of double negation

$$\neg(\neg P) \text{ is equivalent to } P \text{ (i.e. their truth-values are the same)},$$

the commutativity of the conjunction res. disjunction

$$Q \wedge P \text{ is equivalent to } P \wedge Q, \text{ res. } Q \vee P \text{ is equivalent to } P \vee Q,$$

the associativity of the conjunction

$P \wedge (Q \wedge R)$ is equivalent to $(P \wedge Q) \wedge R$,

and the De Morgan law

$$\neg(P \wedge Q) \text{ is equivalent to } (\neg P) \vee (\neg Q),$$

are examples of preserved relations. By contrast, the equivalence of $P \wedge P$ and P is not preserved.

The fuzzy sets F and G, together with the fuzzy implication I, determine a fuzzy relation R in the Cartesian product $\Omega \times \Theta$ described by the membership function

$$\mu_R(x, y) = I(\mu_F(x), \mu_G(y)) \text{ for } x \in \Omega, y \in \Theta.$$

(This is similar to the T-granules defined above.)

From a formal point of view, the chosen fuzzy implication may help us also to define a truth-value of the fuzzy proposition $P \Rightarrow Q$ – where P and Q are fuzzy facts, which is a notation of the following

IF $V(o)$ is F, THEN $W(u)$ is G.

Let us remind that the linguistic variable V (appearing in fuzzy fact P) is a "crisp" function $V : S \to \Omega$, where S is a "crisp" set of objects. Similarly, W is a function $W : S' \to \Theta$, where S' is (possibly) another set of objects. The fuzzy proposition $P \Rightarrow Q$ induces a "crisp" function $f : S \to S'$ such that $u = f(o)$. Therefore, it is enough to consider the following

IF $V(o)$ is F, THEN $W(o)$ is G

or, in general,

IF V is F, THEN W is G.

having in mind that V and W apply to the same object.

Logics is not reduced to a formal iterative construction of formulas, it supposes also obtaining truth-values by inference rules.

Classical logic is based on several unanimously accepted inference rules: *Modus Ponens, Modus Tollens, Syllogismus* etc.

Fuzzy logic is much richer in inference rules as classical logic. Some of these rules are obvious, as for example **the entailment principle**:

from	$V(o)$ is A	Example:	(Age of) John is very-young
and	$\mu_B \geq \mu_A$		$\mu_{young} \geq \mu_{very\text{-}young}$
we infer	$V(o)$ is B		(Age of) John is young

(Here A and B are fuzzy sets in the same universe Ω.)

Another set of obvious inference rules involve fuzzy quantifiers, as for example the following obvious **dispositional rule**:

from	Most $V(o)$ are (is) A	Most Ontarians are very clever
and	$\mu_B \geq \mu_A$	$\mu_{\text{clever}} \geq \mu_{\text{very-clever}}$
we infer	Most $V(o)$ are (is) B	Most Ontarians are clever

The *Generalized Modus Ponens* deserves special attention. In the fuzzy context this rule is presented as follows:

from	$V(o)$ is A
and	IF V is A', THEN W is B'
we infer	$W(o)$ is B

Here A and A' are fuzzy sets in the same universe Ω (of the variable V). On the other side B and B' are fuzzy sets in the universe Θ of the variable W.

As an example, knowing that "(height of) John is TALL" and "if (the height of) a person is VERY TALL, then (the quality of) that person (as a basketball player) is EXCELLENT" we infer "(the quality of) John (as a basketball player) is GOOD".

The *Generalized Modus Ponens* does not require the coincidence of the fuzzy fact "$V(o)$ is A" with the antecedent "$V(o)$ is A'" in the IF-THEN rule. Notice that this is very different from classical logic, which requires that they match exactly.

How this is interpreted? Remember the truth-value of a fuzzy fact "$V(o)$ is A" is exactly $\mu_A(x)$, where x stands for the "exact" value of the variable V for our object o. This truth-value is supposed known.

The truth-value of the conditional proposition

$$\text{IF } V(o) \text{ is } A', \text{ THEN } W(o) \text{ is } B'$$

is expressed as $I(\mu_{A'}(x), \mu_{B'}(y))$, where I is the chosen fuzzy implication. Here y stands for the "exact" value of the variable W for the object o. Of course, this truth-value depends on the object o, and on the fuzzy sets A' and B'.

On the other side, the hypothesis in *Generalized Modus Ponens* is a conjunction of this conditional proposition and of the fuzzy fact

$$V(o) \text{ is } A.$$

Since the latter has truth-value $\mu_A(x)$, for this conjunction the following truth-value is obtained

$$T(\mu_A(x), I(\mu_{A'}(x), \mu_{B'}(y)))$$

where T is the chosen t-norm.

Now, the truth-value of the fuzzy fact "$W(o)$ is B" may be defined as $\mu_B(y)$ once we know the membership function of B. A definition of fuzzy set B is as follows:

$$\mu_B(z) = \sup_{x \in \Omega}\{T(\mu_A(x), I(\mu_{A'}(x), \mu_{B'}(z)))\} \text{ for } z \in \Theta.$$

The simplest situation is obtained when we consider the obvious t-norm $\min\{p, q\}$ and the fuzzy implication $\min\{1, 1 - p + q\}$. Many different choices are possible.

Consider the fuzzy relation R – in the Cartesian product $\Omega \times \Theta$ – determined by the fuzzy sets A' and B' and by the fuzzy implication I. Then the formula above defines B as a kind of composition $A \circ R$. Of course, $B = A \circ R$ in case $T = \min$.

The *Generalized Modus Ponens*, as presented above, supposes the fuzzy sets A, A' and B' are defined. It helps to describe a new fuzzy set B. However, the construction above produces rather imprecise fuzzy sets B, even in the simplest situation.

In Figure 9.19 such an example is presented. Here $\Omega = \Theta = \mathbf{R}$ and A is a translate of A'. The t-norm is $\min\{p, q\}$ and the fuzzy implication is $\min\{1, 1 - p + q\}$. Not only the core of B is larger than the core of B', but our impression is that B is not so "near" to B' at the extent A is to A'.

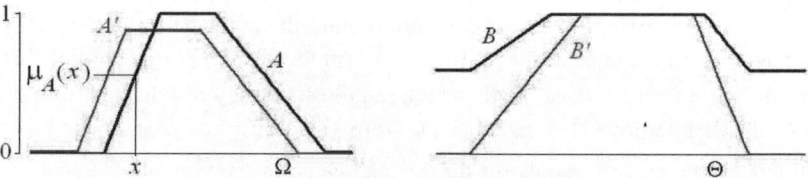

Fig. 9.19. Results of GMP

In what follows a measure of consistency of two fuzzy sets in the same universe is needed. Such a measure is given, once a t-norm T is chosen, by the T-**degree of consistency**

$$d(A, A') = \sup_{x \in \Omega}\{T(\mu_A(x), \mu_{A'}(x))\} \text{ for } A, A' \text{ fuzzy sets in } \Omega.$$

The **approximate reasoning** is based on suitable choices of fuzzy sets $A_1, A_2, ..., A_n$ in the universe Ω of the variable V, of fuzzy sets $B_1, B_2, ..., B_n$ in the universe Θ of the variable W, and on the general schema:

> IF $V(o)$ is A_1, THEN $W(o)$ is B_1
>
> IF $V(o)$ is A_2, THEN $W(o)$ is B_2
>
> ...
>
> IF $V(o)$ is A_n, THEN $W(o)$ is B_n
>
> $V(o)$ is A
> _____
>
> $W(o)$ is B

The fuzzy sets $A_1, A_2, ..., A_n$ res. $B_1, B_2, ..., B_n$ in most cases are chosen of triangular type, as for example the following:

NL (negative large)	PS (positive small)
NM (negative medium)	PM (positive medium)
NS (negative small)	PL (positive large)
	Z (approx. zero)

to cover main variation interval $[-a, a]$ as in the Figure 9.20 below:

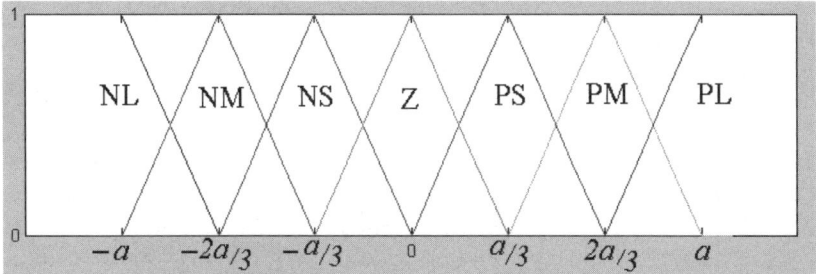

Fig. 9.20. Triangular fuzzy sets covering an interval

The approximate reasoning is an extension of the well-known polynomial interpolation problem: given n pairs of real or complex numbers (x_1, y_1), (x_2, y_2), ..., (x_n, y_n) with $x_k \neq x_j$ for $k \neq j$ – called nodes (or support points, see [Stoer and Bulirsch 1980]), the Lagrange formula

$$L(x) = \sum_{j=1}^{n} y_j \prod_{k \neq j} \frac{x - x_k}{x_j - x_k}$$

gives the value in x of the polynomial L that interpolates through all nodes (see Figure 9.21a). This formula involves a "degree of consistency" $\prod_{k \neq j} \dfrac{x - x_k}{x_j - x_k}$ between the number x and the support abscissa x_j for $j \in \{1, 2, ..., n\}$.

In case Ω and Θ are contained in $\mathbf{R}$ and are respectively covered by seven fuzzy sets as in Figure 9.21b, then the analogue of the interpolation polynomial is formed by seven (min-)granules.

The approximate reasoning is typical in fuzzy logic controllers. The most common way to determine fuzzy set B is referred to as the **method of interpolation**. It consists of the following two steps:

Step 1. Calculate the degree of consistency r between the given fuzzy fact "$V(o)$ is A" and the antecedent of each IF-THEN rule, for example using the formula:

$$r_j = \sup_{x \in \Omega} \{\min \{\mu_A(x), \mu_{A_j}(x)\}\}, \quad j \in \{1, 2, ..., n\}.$$

Step 2. Calculate the membership function of the fuzzy set B by truncating each fuzzy set B_j at level r_j and then taking the union of the truncated sets:

$$\mu_B(y) = \max_j \{\min \{r_j, \mu_{B_j}(y)\}\}.$$

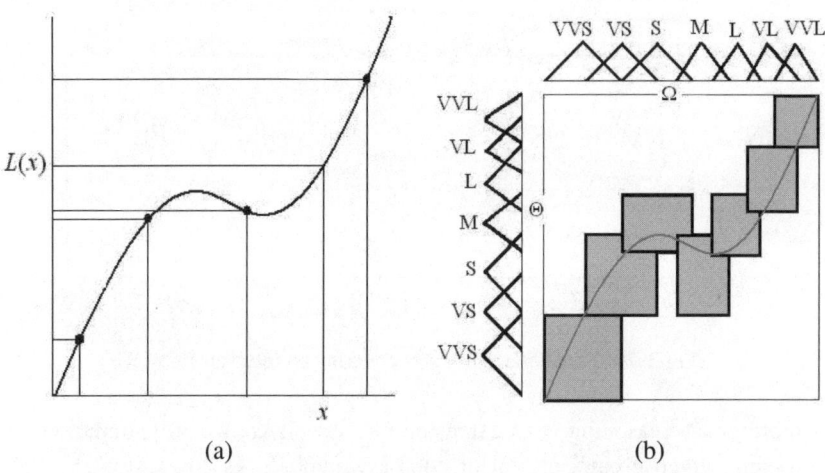

Fig. 9.21. Interpolation in $\mathbf{R}^2$ (a) and fuzzy rules (b)

An example of interpolation is presented in Figure 9.22.

9.6 Defuzzification

Solving some problems supposes the final result should be expressed in crisp terms (e.g. as a real number). The result of a fuzzy inference process (or of a fuzzy approximate reasoning) is a fuzzy set. Replacing this fuzzy set by a suitable crisp value is known as **defuzzification**.

Several methods are available.

1) The **center of gravity method** supposes that the membership value $\mu_B(y)$ is a kind of weight attached to the value y from the universe. A weighted mean is then considered

$$\bar{y} = \frac{\sum \mu_B(y) y}{\sum \mu_B(y)} \left(\text{or} \; \frac{\int_Y \mu_B(y) y \, dy}{\int_Y \mu_B(y) \, dy} \right)$$

and this crisp value is chosen as final result.

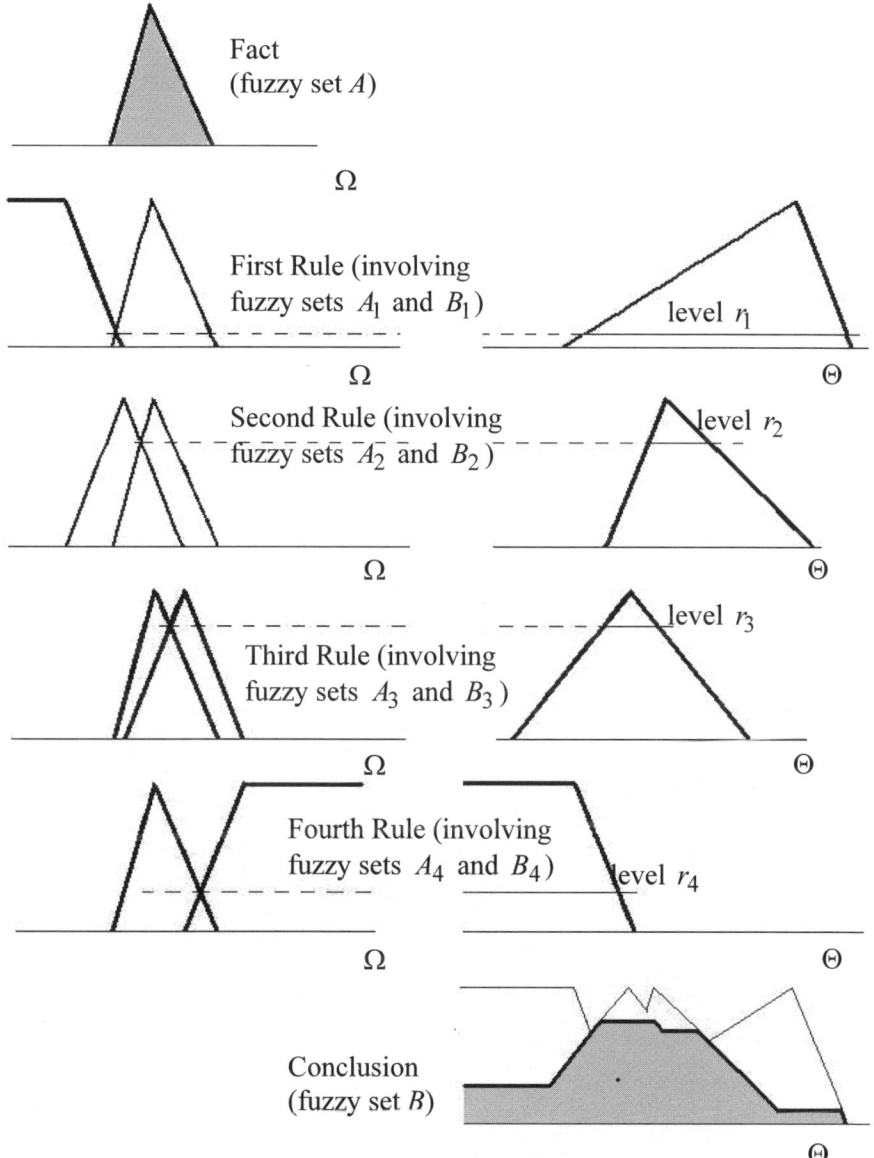

Fact
(fuzzy set A)

Ω

First Rule (involving
fuzzy sets A_1 and B_1)

level r_1

Ω

Θ

Second Rule (involving
fuzzy sets A_2 and B_2)

level r_2

Ω

Θ

level r_3

Third Rule (involving
fuzzy sets A_3 and B_3)

Ω

Θ

Fourth Rule (involving
fuzzy sets A_4 and B_4)

level r_4

Ω

Θ

Conclusion
(fuzzy set B)

Θ

Fig. 9.22. Example of interpolation

This method is implemented in function **defuzz** in *Matlab*, as option **centroid**. It works well in case of a continuum universe Θ covered by triangular fuzzy sets. When the universe Θ is discrete it is very possible that the obtained value $\bar{y}$ to be outside the universe, thus not a genuine value. In the example above we have

$$\bar{y} = \frac{0.2y_1 + 0.3y_2}{0.2 + 0.3} = 0.4y_1 + 0.6y_2$$

which is neither y_1 nor y_2 !

2) The **center of maxima method** takes into account only the elements $y \in \Theta$ where the membership values to the resulted fuzzy set B are maximal (maybe < 1). Thus the set

$$M = \{ y \mid \mu_B(y) \text{ is maximal} \}$$

is considered. In a "tame enough" universe Θ this set M has a supremum and an in-fimum. Their arithmetic mean

$$m = \frac{1}{2}(\sup M + \inf M)$$

is chosen as the final result.

In the example above the set M is limited to the singleton set $\{y_2\}$, thus $m = y_2$ is the final result of the defuzzification process.

This method works well especially when M is a connected set. When M is discon-nected, another method could be imagined.

3) The **mean of maxima method** (implemented as option **mom** in **defuzz**). Now a kind of mean value $\bar{m}$ of elements of the set M above is considered.

In the Figure 9.23 a comparison between the crisp values obtained by these three methods is presented, supposing that the fuzzy set B is known.

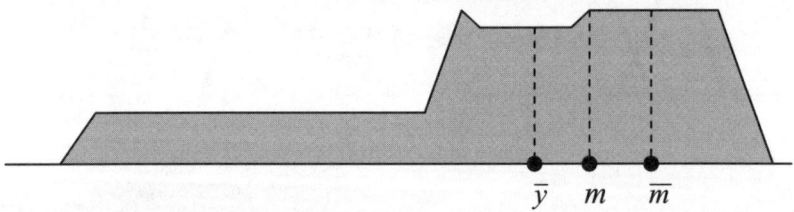

$\bar{y} \quad m \quad \bar{m}$

Fig. 9.23. Results of defuzzification

9.7 Approach by Precision Degrees

Another approach takes into account the notion of precision degree of a proposition. Sometimes a sentence can be decomposed into a first part expressing an idea (or a fact) and a second part expressing the imprecision of the first part. For example, in the sentence "It is almost sure that in a flu case the temperature of patient is quite high" the imprecision is expressed by "it is almost sure".

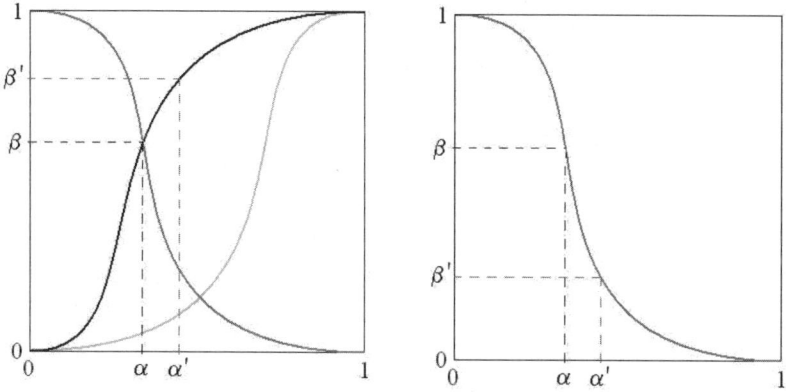

Fig. 9.24. Precision degree β' given by Bézier curves: Case 1 (left) vs. Case 2 (right)

In general, such imprecise sentences contain a fact "X is A" and an attached precision degree $\alpha \in [0,1]$. From a formal point of view, these sentences could be expressed as "X is αA".

Using precision degrees, the *Generalized Modus Ponens* can be expressed as follows:

from	IF V is αA, THEN W is βB
and	$V(o)$ is $\alpha' A$
we infer	$W(o)$ is $\beta' B$

It is worth noting here that for a pair A, B of fuzzy sets two distinct cases are identified:

Case 1. When the degree of precision of "V is A" is raised from α to α', the degree of precision of "W is B" also raises from β to a higher value β'.

(Usually in this case the fuzzy sets are chosen such that for $\alpha = 1$ one has $\beta = 1$ and for $\alpha = 0$ one has $\beta = 0$.)

Case 2. When the degree of precision of "V is A" is raised from α to a higher α', the degree of precision of "W is B" decreases from β to a lower value β'.

The degree β' in the *Generalized Modus Ponens* rule is obtained from a formula involving the three other degrees of precision

$$\beta' = \varphi(\alpha, \beta, \alpha').$$

Following [Khoukhi 1996], the inference process consists of:

a) computing the similarity degree between the precision degree α' of the fact and the precision degree α of the antecedent

$$\sigma = \min\{I(\alpha, \alpha'), I(\alpha', \alpha)\},$$

where I is the chosen fuzzy inference, and then

b) determining the precision degree β' by taking into account the above similarity degree, for example as follows:

$$\beta' = \max\{0, \beta + \sigma - 1\}.$$

In Figure 9.24 the function φ is described by families of Bézier curves, indexed by a parameter $\lambda \in [0,1]$, covering the unit square. For example, in Case 1 the end nodes are $A = (0,0)$ and $B = (1,1)$. The control nodes are $C = (1,0)$ and $D = (\lambda,1)$, or $C = (\lambda,0)$ and $D = (0,1)$.

9.8 Solved Exercises

1) Assume "young (age)" is represented by a fuzzy set Y

$$\mu_Y(x) = \begin{cases} 1 & x \le 20 \\ (40 - x)/20 & 20 < x < 40 \\ 0 & x \ge 40 \end{cases}$$

and "old (age) is represented by

$$\mu_O(x) = \begin{cases} 0 & x \le 50 \\ (x - 70)/20 & 50 < x < 70 \\ 1 & x \ge 70 \end{cases}$$

Identify fuzzy term "neither young, nor old" (i.e. "middle aged"). Show that "old" implies "not young".

2) Given the fuzzy sets – in the universe $X = \{1,2,3,4,5\}$

$A = \{(0.49|1), (0.64|2), (1|3), (0.16|4), (0.81|5)\}$ and
$B = \{(0.7|1), (0.6|2), (1|3), (0.9|4), (0|5)\}$,

calculate the following

    (a) the union $A \cup B$,
    (b) the algebraic sum $A + B$,
    (c) the Cartesian product $A \times B$,
    (d) more or less A,
    (e) more or less A and very B.

3) Assuming FALSE is a fuzzy set over the universe $[0,1]$, described by a Z-type membership function, represent the fuzzy set

    (a) SOMEWHAT FALSE

  (b) FALSE AND NOT FALSE
  (c) SOMEWHAT FALSE AND NOT FALSE

4) Assuming TRUE is a fuzzy set over the universe [0, 1], described by an S-type membership function, draw the fuzzy set

(a) NOT TRUE   (b) SOMEWHAT TRUE

(c) TRUE AND NOT TRUE   (d) SOMEWHAT TRUE AND NOT TRUE

5) The IQ of an individual is a number around 100. (In probabilistic terms, it is obtained from a normal (Gaussian) random variable with mean 100 and standard deviation 15.)

Describe a fuzzy variable based on IQ values, but having fuzzy values "smart", "very smart", "rather smart". Usually 16% of the individuals are considered smart.

6) By the normalization procedure, from a fuzzy set A one obtains the fuzzy set NORM(A) – over the same universe X – whose membership function is

$$\mu_{\text{NORM}(A)}(x) = \frac{1}{\max\limits_{z \in X} \mu_A(z)} \mu_A(x).$$

By the intensification procedure, from a fuzzy set A one obtains the fuzzy set INTENS(A) – over the same universe X – whose membership function is

$$\mu_{\text{INTENS}(A)}(x) = \begin{cases} 2(\mu_A(x))^2 & \text{if } \mu_A(x) \leq 0.5 \\ 1 - 2(1 - \mu_A(x))^2 & \text{if } \mu_A(x) > 0.5. \end{cases}$$

Knowing that "SLIGHTLY A" is defined as
    INTENS(NORM(A and NOT(very A)))
give a complete definition of the fuzzy sets SLIGHTLY F, SLIGHTLY G and SLIGHTLY $F \vee$ SLIGHTLY G, where:

   F = TALL = trapmf(x, 150, 190, ∞, ∞)   (trapezoidal)

   G = MEDIUM = trimf(x, 130, 160, 190)   (triangular)

7) Denote $T : [0,1] \times [0,1] \rightarrow [0,1]$ the function defined by $T(0,0) = 0$ and

$T(p,q) = \dfrac{p \cdot q}{p + q - p \cdot q}$ if $p \cdot q \neq 0$.

(a) Show that T is a t-norm.
(b) Show that T is intermediate between the t-norms product and min, i.e. $p \cdot q \leq T(p,q) \leq \min\{p,q\}$.

8) The "middle point" $P(\frac{1}{2})$ of a Bézier curve determined by end nodes A, B and control nodes C, D is obtained as follows:

(a) Construct the middle point C' of the segment AC, the middle point E of CD and the middle point D'' of BD,

(b) Construct the middle point D' of the segment $C'E$, and the middle point C'' of the segment $D''E$,

(c) Construct the middle point X of the segment $C''D'$. This is exactly $P(\frac{1}{2})$.

Show that the Bézier curve determined by A and X as end nodes and C', D' as control nodes is exactly the "first half" of the original Bézier curve. The second half is the Bézier curve determined by X and B as end nodes and C'', D'' as control nodes.

9) The fuzzy set B (in the universe $\{1, ..., 9\}$ is defined as follows:

$$0.1|1 + 0.9|2 + 0.6|3 + 0.1|4 + 0.4|5 + 0.9|6 + 0.8|7 + 0.7|8 + 0.5|9$$

Defuzzify this set using:

a) the center of maxima method;
b) the center of gravity method.

10) The **generalized *modus tollens*** is an inference rule expressed as follows:

$$\text{IF } V \text{ is } A, \text{ THEN } W \text{ is } B$$
$$\underline{W(o) \text{ is } B'}$$
$$V(o) \text{ is } A'$$

Derive a formula to compute the fuzzy set A' given fuzzy sets A, B and B'.

11) The **sigma-count** and the **sigma-mean** of a fuzzy set A over a finite universe U are defined via the membership function μ_A as follows:

$$\Sigma\#(A) = \sum_{u \in U} \mu_A(u), \text{ res. } \Sigma M(A) = \frac{\Sigma\#(A)}{\Sigma\#(U)}.$$

Show that

$$\Sigma\#(A \cap B) + \Sigma\#(A \cup B) = \Sigma\#(A) + \Sigma\#(B)$$

and

$$\max(0, \Sigma M(A) + \Sigma M(B) - 1) \le \Sigma M(A \cap B) \le \min(\Sigma M(A), \Sigma M(B))$$

for any fuzzy sets A and B in U.

Solutions

1) Fuzzy set $M =$ "neither young, nor old" is the intersection $\overline{Y} \cap \overline{O}$, i.e. the trapezoidal fuzzy set with $ld = 20$, $lu = 40$, $ru = 50$, $rd = 70$. It is immediate that $\mu_O(x) \le \mu_{\overline{Y}}(x)$ for all x, i.e. $O \le \overline{Y}$.

2) Technical exercise.

3) Suppose FALSE is the trapezoidal fuzzy set trapmf(x, 0, 0, 0, 0.5). In this case FALSE AND NOT FALSE is $\dfrac{1}{2}\operatorname{trapmf}(x, 0, 0.25, 0.25, 0.5)$, a non-normal fuzzy set.

4) TRUE AND NOT TRUE is also a non-normal fuzzy set.

5) Identify first the 94th percentile of the normal distribution with mean 100 and variance $15^2 = 225$. By use of function NORMINV in Microsoft Excel, this percentile is found at about 123. A very simple description of "smart" could be given by $\operatorname{trapmf}(x, 100, 123, \infty, \infty)$.

(Remark. No negative IQ values exist. Therefore, to be absolute correct, we should truncate at 0 the normal distribution above. However, the probability of negative values is 0.000000000013, a very small value that can be neglected.)

7) (a) Conditions (T0) and (T1) are obvious. All the three conditions (T2)-(T3)-(T4) of t-norms are easy to be verified if we notice that $T(p, q) = \varphi^{-1}(\varphi(p) + \varphi(q))$ where $\varphi : (0, 1] \to [0, \infty)$ is the one-to-one correspondence given by $\varphi(p) = \dfrac{1}{p} - 1$.

8) It is easy to obtain $C' = \frac{1}{2}A + \frac{1}{2}C$, $D' = \frac{1}{4}A + \frac{1}{2}C + \frac{1}{4}D$ etc. The point X is expressed as $\frac{1}{8}A + \frac{1}{8}B + \frac{3}{8}C + \frac{3}{8}D$, i.e. as $P(\frac{1}{2})$.

If the Bézier curve determined by A, B, C, D is described by the cubic polynomial $P(t) = (1 - t)^3 A + 3t(1 - t)^2 C + 3t^2(1 - t)D + t^3 B$, then the Bézier curve determined by A, X, C', D' is described by the cubic polynomial $Q(s) = (1 - t)^3 A + 3t(1 - t)^2 C + 3t^2(1 - t)D + t^3 B$, i.e. $Q(s) = P(2t)$.

9) The fuzzy set B is not normal. The maximal membership values are attained at 2 and at 6. The result of the defuzzification by center of maxima method is thus 4, which has minimal membership value!

On the other hand, the center of gravity method gives 5.44, which is not a member of the universe!

10) Consider the complements $\neg A$, $\neg B$ of fuzzy sets A, B. Replace the "knowledge"

IF V is A, THEN W is B

by its contraposition

IF W is $\neg B$, THEN V is $\neg A$,

From "$W(o)$ is B'" with known fuzzy set B', applying the Generalized Modus Ponens we infer "V is A'" where the membership function of fuzzy set A' is computed as follows

$$\mu_{A'}(x) = \sup_y T(\mu_{B'}(y), I(\mu_{\neg B}(y), \mu_{\neg A}(x))).$$

If the chosen fuzzy implication I satisfies (I5) for the classical negation, then the fuzzy set A' is obtained from:

$$\mu_{A'}(x) = \sup_{y} T(\mu_{B'}(y), I(\mu_A(x), \mu_B(y))).$$

11) The relation

$$\Sigma\#(A \cap B) + \Sigma\#(A \cup B) = \Sigma\#(A) + \Sigma\#(B)$$

is obvious, since $\mu_{A \cap B}(u) = \min\{\mu_A(u), \mu_B(u)\}$ for any fuzzy sets A and B in U. Notice that the sigma-count satisfies the monotonic condition $\Sigma\#(A) \le \Sigma\#(B)$ if $A \subseteq B$.

Starting from the sigma-count, by analogy with probabilities, a conditional sigma-count is defined by

$$\Sigma\#(A \mid B) = \frac{\Sigma\#(A \cap B)}{\Sigma\#(B)}$$

(for all non-trivial fuzzy sets B).

10 Review

10.1 Review of Uncertainty and Imprecision

Consider the unit interval [0, 1]. Its elements may be interpreted as:

- Probabilities of events,
- Possibilities of propositions,
- Beliefs of propositions,
- Membership degrees of elements,
- Truth values of propositions, etc.

and all kind of confusions may appear in the natural language.

Progress in Artificial Intelligence is not possible without defining **units** for measuring uncertainty and imprecision. Smets [Smc 2000] suggested calling these units by adding the suffix "it" (from "information unit"). Expressed in these units, all the values should be between 0 and 1.

Thus:

1) **probit** is the unit for probabilities. Instead of writing

P(the house of Sally is burgled) = 0.2

we may write now "the house of Sally is burgled" has probit 0.2;

2) **possit** is the unit for possibilities. Instead of

Π(John is coming) = 1
Π(John is not coming) = 0.1

(i.e. "John is coming almost sure") we may write now "John is coming" has possit 0.95.

The transformation formula involves both possibilities:

$$possit \ value = \frac{\Pi(p) + 1 - \Pi(\neg p)}{2};$$

3) **belit** is the unit for beliefs. Instead of

$$\text{Bel}(r \vee f) = 0.2$$

we may write now "The paint is a Raphael or a forgery" has belit 0.2;

E. Roventa and T. Spircu: Management of Knowledge Imperfection, STUDFUZZ 227, pp. 233–246.
springerlink.com

4) **fuzzit** is the unit for membership grades. Instead of

$$\mu_{TALL}(180) = 0.6$$

we may write now "180 cm is tall" has fuzzit 0.6;

5) **verit** is the unit for truth-values. Instead of

"most tall men are not very fat"

we may write now "tall men are not very fat" has verit 0.8.

Consider, in the following, the universe of discourse Ω. This has:

- elements ω,
- crisp subsets A, forming the set $\mathcal{P}(\Omega) = 2^{\Omega}$,
- fuzzy subsets F, forming the set $\mathcal{F}(\Omega)$.

Let us review the measures used in uncertain information processing.

1) A **probability distribution** over Ω (finite) is described as a two-rows table

$$\begin{matrix} \text{elements} \\ \text{probits} \end{matrix} \begin{pmatrix} \dots & \omega & \dots \\ \dots p(\omega) \dots \end{pmatrix}$$

where the sum of all probits $p(\omega)$ is equal to 1. The function p is extended by

$$P(A) = \sum_{\omega \in A} p(\omega)$$

to the crisp subsets A of Ω and satisfies the **additivity relation**

$$P(A \cup B) = P(A) + P(B) - P(A \cap B).$$

2) A **possibility distribution** over Ω (finite) is described as a two-rows table

$$\begin{matrix} \text{elements} \\ \text{possits} \end{matrix} \begin{pmatrix} \dots & \omega & \dots \\ \dots \pi(\omega) \dots \end{pmatrix}$$

(the "sum equal to 1" condition is not required here). The function π is extended by

$$\Pi(A) = \max_{\omega \in A} \pi(\omega)$$

to the crisp subsets A of Ω and satisfies the **max relation**

$$\Pi(A \cup B) = \max(\Pi(A), \Pi(B)).$$

3) A **basic belief assignment** over Ω is described as a two-rows table

$$\begin{matrix} \text{(crisp) subsets} \\ \text{belits} \end{matrix} \begin{pmatrix} \dots & A & \dots \\ \dots \text{Bel}(A) \dots \end{pmatrix}$$

where $\varnothing$ has belit 0 and Ω has belit 1, or as a two-rows table

$$\begin{matrix} \text{subsets} \\ \text{mass distribution} \end{matrix} \begin{pmatrix} ... & A & ... \\ ... & m(A) & ... \end{pmatrix}$$

where the masses $m(A)$ sum to 1. Remember the relations between the belits and mass distribution:

$$\text{Bel}(A) = \sum_{B \subseteq A} m(B) \text{ and } m(A) = \sum_{B \subseteq A} (-1)^{|A-B|} \text{Bel}(B).$$

The belief assignment function satisfies the **super-additivity relation**:

$$\text{Bel}(A \cup B) \geq \text{Bel}(A) + \text{Bel}(B) - \text{Bel}(A \cap B).$$

4) A **fuzzy set** in Ω can be described as a two-rows table

$$\begin{matrix} \text{elements} \\ \text{fuzzits} \end{matrix} \begin{pmatrix} ... & \omega & ... \\ ... & \mu_F(\omega) & ... \end{pmatrix}$$

without any restriction.

By combining fuzziness with the measures 1)-3) above we obtain several measures of fuzzy sets. All these should satisfy an obvious requirement, called *monotonicity*:

(M)   If $\mu_F \geq \mu_G$ for fuzzy sets F and G, then the measure of G is at least the measure of F.

Let us consider some examples.

5) A **probability distribution over fuzzy sets** in Ω is described as a two-rows table

$$\begin{matrix} \text{fuzzy sets} \\ \text{probits} \end{matrix} \begin{pmatrix} ... & F & ... \\ ... & p(F) & ... \end{pmatrix}$$

where, as usual for probability distributions, the sum of all probits $p(F)$ is equal to 1. (Here we accept that the number of distinct fuzzy membership values is finite.)

6) A **possibility distribution over fuzzy sets** in Ω is described as a two-rows table

$$\begin{matrix} \text{fuzzy sets} \\ \text{possits} \end{matrix} \begin{pmatrix} ... & F & ... \\ ... & \Pi(F) & ... \end{pmatrix}$$

with no other restriction on the possits.

7) A **basic belief assignment over fuzzy sets** in Ω is described as a two-rows table

$$\text{fuzzy sets} \begin{pmatrix} \dots & F & \dots \\ \dots \text{Bel}(F) \dots \end{pmatrix}$$

with no other restriction on the belits.

By combining fuzziness with the measures above we may obtain also fuzzy measures. Before defining this concept, let us consider only one example.

8) A **fuzzy probability distribution** over Ω (finite) is described as a two-rows table

$$\text{elements} \begin{pmatrix} \dots & \omega & \dots \\ \dots FP(\omega) \dots \end{pmatrix}$$
$$\text{fuzzy probits}$$

where each $FP(\omega)$ is a fuzzy set in $[0, 1]$. Consider $FP(\omega)$ is defuzzified by the center of gravity method to the crisp value $c(\omega)$; then the following condition should be satisfied:

$$\sum_{\omega} c(\omega) = 1 .$$

Let us treat the notions above from the point of view of fuzzy measures. Consider a family C of subsets of the universe Ω, satisfying the following lattice conditions

(C1) $\varnothing \in C$ and $\Omega \in C$,

(C2)   If $A, B \in C$, then $A \cup B \in C$ and $A \cap B \in C$.

A **fuzzy measure** on C is a function

$$g : C \rightarrow [0, 1]$$

satisfying the following two conditions:

(FM1) $g(\varnothing) = 0$, $g(\Omega) = 1$ (boundary conditions),

(FM2)   If $A, B \in C$ and $A \subseteq B$, then $g(A) \leq g(B)$ (g is monotonic).

The number $g(A)$ assigned to a subset $A \in C$ is interpreted as a measure of the total available evidence that a given element of Ω, whose characterization is deficient in some way, belongs to the subset A.

Example. Suppose we are trying to diagnose an ill patient. We may be trying to determine whether this patient belongs to the set Ω of people with pneumonia, bronchitis, emphysema, or just a common cold. We may consider that the subsets in C are exactly these six:

$\varnothing, A = \{\text{pneumonia}\}, B = \{\text{pneumonia, emphysema}\},$

$C = \{\text{pneumonia, bronchitis}\}, D = \{\text{common cold}\}$ and Ω.

Now, in defining a fuzzy measure g on C we have to respect the two strict inclusions $A \subset B$ and $A \subset C$ and we may define $g(D)$ independently of the other values of g.

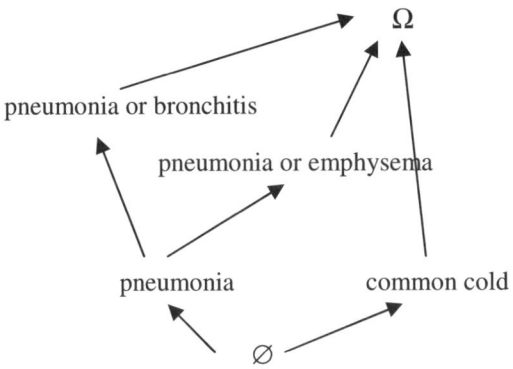

In general, given a fuzzy measure g on C and $A, B \in C$, because of the inclusions $A \cap B \subseteq A$ and $A \cap B \subseteq B$, the monotonic condition allows us to infer the following inequality:

$$g(A \cap B) \leq \min\{g(A), g(B)\}.$$

Dually, because of the inclusions $A \subseteq A \cap B$ and $B \subseteq A \cap B$, the same monotonic condition allows us to infer another inequality:

$$g(A \cup B) \geq \max\{g(A), g(B)\}.$$

Consider now some important particular cases.

a) First, in the case of a probability distribution over a finite set Ω, the family C is exactly the set $\mathcal{P}(\Omega)$ of all subsets of Ω, and the function $\mathcal{P}$ plays the role of g.

(It is obvious that $P(\Omega) = 1$. If $A, B \in \mathcal{P}(\Omega)$ and $A \subseteq B$, then $P(B) = P(A \cup (B - A)) = P(A) + P(B - A) \geq P(A)$ since $A \cap (B - A) = \varnothing$, thus $P(A) \leq P(B)$. Therefore $P(A \cap B) \leq \min\{P(A), P(B)\}$ and $P(A \cup B) \geq \max\{P(A), P(B)\}$.

b) In the case of the belief assignment, the family C is the same set $\mathcal{P}(\Omega)$ of all subsets of Ω, now the function Bel plays the role of g.

(It is very easy to show that if $A, B \in \mathcal{P}(\Omega)$ and $A \subseteq B$, then $\mathrm{Bel}(B) = \mathrm{Bel}(A \cup (B - A)) \geq \mathrm{Bel}(A) + \mathrm{Bel}(B - A) - \mathrm{Bel}(A \cap (B - A)) \geq \mathrm{Bel}(A)$, because $\mathrm{Bel}(A \cap (B - A)) = \mathrm{Bel}(\varnothing) = 0$, thus $\mathrm{Bel}(A) \leq \mathrm{Bel}(B)$. It follows that $\mathrm{Bel}(A \cap B) \leq \min\{\mathrm{Bel}(A), \mathrm{Bel}(B)\}$ and $\mathrm{Bel}(A \cup B) \geq \max\{\mathrm{Bel}(A), \mathrm{Bel}(B)\}$, but these inequalities may be strict even in the absence of a particular relation between A and B.)

If we force a belief assignment Bel to be a probability measure, then the associated basic belief assignment m should have value 0 on each subset of Ω that is not a singleton.

This result makes a clear distinction between the two particular (fuzzy) measures: a probability measure is just a particular case of a basic belief assignment.

Remember how the total ignorance is expressed by a belief assignment: all subsets A different from Ω have belit 0 (and, of course, $\mathrm{Bel}(\Omega)=1$).

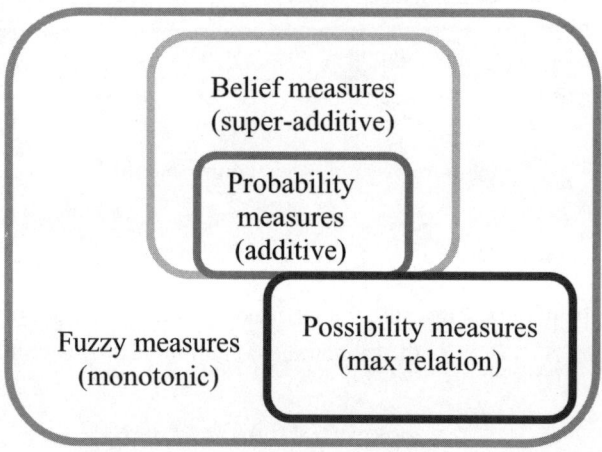

Fig. 10.1. Fuzzy measures

c) In the case of the possibility distribution over a finite set Ω, the family C is again the set $\mathcal{P}(\Omega)$ of all subsets of Ω, and now the function Π plays the role of g.

(Now if $A,B \in \mathcal{P}(\Omega)$ and $A \subseteq B$, then $A \cup B = B$. From the condition $\Pi(A \cup B) = \max\{\Pi(A), \Pi(B)\}$ it follows immediately $\Pi(A) \le \Pi(B)$.)

At a first glance, the possibility measure and the probability measure seem to be analogous, because of the similarity between the definition formulas $P(A) = \sum_{\omega \in A} p(\omega)$ res. $\Pi(A) = \max_{\omega \in A} \pi(\omega)$. However, the additivity relation $P(A \cup B) = P(A) + P(B) - P(A \cap B)$ is not analogous to the max relation $\Pi(A \cup B) = \max\{\Pi(A), \Pi(B)\}$. Another difference between these measures is in the manner total ignorance is expressed. In the case of possibility measures the total uncertainty is expressed by the fact that all elementary possits $\pi(x)$ are equal to 1, which implies $\Pi(A) = 1$ for all subsets A of Ω. On the contrary, in the case of probability measures the total uncertainty is expressed by the equality of all probits $p(\omega)$.

In the Figure 10.1 above, the relations between these measures, as particular cases of fuzzy measures are presented.

10.2   Production Rules

Let us consider some examples of IF-THEN rules that may appear in an expert system.

(1)   IF fuel tank is empty
      THEN engine is dead

This IF-THEN rule expresses a simple relation. The rule is "certain", no doubt; what about its antecedent and its consequent?

Of course, "dead" is a binary value of a Boolean variable, the state of the engine. The other value is "running". Thus, the consequent of the rule could be considered as being precise.

On the contrary, the antecedent of the rule above is imprecise, and the imprecision is dealt with in the context of fuzzy sets.

We recognize in the antecedent that "empty" is a fuzzy set. In fact, we have a variable (the capacity of the fuel tank), which is of linguistic type. Thus we may consider an imprecise value fuzzit(empty). Since our rule is certain, i.e. belit(rule) = 1, this value is transferred automatically into verit(engine is dead) that is transformed into a crisp truth-value.

On the other side, the values of the capacity of the fuel tank: empty, low, medium, nearly full, full a.o. may be considered as fuzzy sets, covering the real interval [0, MAX_CAPACITY]. Of course, the upper limit of our interval depends on the type of the car (but, in fact, only on the type of the fuel tank!).

Both "fuel tank" and "engine" may be considered as slots in the frame "car". Another slot is "type".

(2)   IF infection is meningitis
         AND patient is a child
      THEN drug recommendation is ampicilline

This is a classical recommendation. Here several imperfect data appear.

First of all, we have to point out that the recommendation is not mandatory; many physicians recommend other drug treatments in meningitis infections. Thus a degree of belief in our rule, as a whole, appears and this degree of belief – or credibility – is dealt within belief theory. In this particular case, knowing that there exists a similar rule (of the same kind) in which "gentamycin" replaces "ampicilline", when lacking any information about how often the rules are used, we may consider that belit(rule) = 0.5.

The antecedent is a conjunction, and the first term of it (infection is meningitis) usually is subjectively asserted. The best treatment of this kind of assertion is in belief theory, where a degree of belief is attached, as for example belit(infection is meningitis) = 0.95.

The second term of the antecedent is of fuzzy type. It is clear that "child" is a fuzzy set, value of the linguistic variable "age", and a truth-value – derived from a membership degree fuzzit(child) – is obtained for this term.

Now we have to combine a belit value with a fuzzit value, in order to derive a truth-value for the antecedent. One suggestion would be the use of a suitable t-norm:

verit(antecedent) = T(belit(infection is meningitis), fuzzit(child))

and the most suited here seems to be $T(\alpha, \beta) = \alpha \cdot \beta$.

(3)   IF gender is female

    AND age is old
    AND ft4_value is decreased
    AND tsh_value is markedly elevated
THEN (h) diagnosis is overt hypothyroidism
    AND (a) advice is treatment may benefit

This rule is a typical diagnosis-recommendation, very similar to a diagnosis-directive rule, in which "advice" is replaced by "action".

The antecedent is a multiple conjunction, in which the first term ("gender is female") is of Boolean type and all the other three terms are of fuzzy type. It is true, however, that in today's medical practice the values "decreased" for the Free Thyroxine test (FT4) and "markedly elevated" for the Thyroid-Stimulating Hormone test (TSH) are crisp labels determined by the upper bound 0.8 nanograms per deciliter res. by the lower bound 10 milliunits per liter. In the same line of thought "old" for age could mean a crisp label determined by the lower bound 50 years.

Notice the consequent of the rule is a conjunction too. Its first term h is "believed", thus has a degree of belief. On the contrary, a possibility value is attached to the second term (expressing how much the patient may benefit from the standard treatment). Thus in the consequent a belit value is combined with a possit value. (The choice of specific values is a matter of personal choice of the physician. However, these values should be ascertained by statistical evidence/data. For example, well-designed prospective clinical and epidemiologic studies have found that 13 in 1000 women older than 50 have unsuspected but symptomatic overt hypothyroidism or overt hyperthyroidism that will respond to treatment.) For the combination (AND) one suggestion could be to use a t-norm.

Concerning the imprecision attached to the entire rule the following alternative could be suggested:

– either to consider that the rule is certain (i.e. belit = 1), and the degree of imprecision of the consequent is computed following the scheme

    belit(consequent) = T(belit(h), possit(a)),

where T is a (previously chosen) t-norm;

– or to consider that the rule has belit(h) as imprecision value, i.e. that the belit value of the diagnosis is transferred to the rule.

The advantage of the imprecise processing of the rule versus the crisp one becomes obvious if we notice that for a woman of 49, in the crisp case the rule is not fired at all!

Such kinds of rules are common in screening for diseases. Although how the fuzzy set OLD should be defined is known, the definition of the other fuzzy sets involved in this rule should be done following the specific medical guidelines. For example, the values of the variable *tsh_value* could be the following three (trapezoidal) fuzzy sets:

NORMAL = trapmf(0, 0.5, 4.5, 6),
MILDLY_ELEVATED = trapmf(4.5, 6, 9, 10),
MARKEDLY_ELEVATED = trapmf(9, 10, ∞, ∞).

(4) IF infection after wound on foot   IF infection after wound on foot
        AND antecedent is diabetes   AND antecedent is diabetes
        AND age is very old   AND age is very old
      THEN medication and wait   THEN immediate amputation below the
                                      knee

    IF extension of infection
    THEN late amputation above the knee

These three rules describe two different strategies a physician could select. Of course, our physician believes that one is better than the other; it is a matter of personal decision! His belief is based on utility calculus done beforehand.

Suppose belit(left strategy) > belit(right strategy), thus the left strategy (more risky!) is chosen. This involves two successive IF-THEN rules, separated in time. Time is an important factor that should be taken into account.

Of course, the problem of computing beliefs may be approached in terms of Bayesian networks, based on the (simplified) graph in Figure 10.2 nd on the estimation of all necessary conditional probabilities.

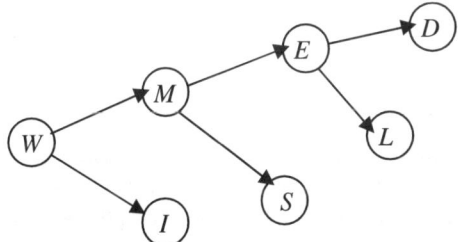

W – wound on foot
M – medication and wait
I – immediate amputation
E – extension of infection
S – scar formation (healing)
D – death
L – late amputation

Fig. 10.2. Graph of possible evolution of an infection

Of course, the value of interest here is $p(l \mid m)$, i.e. the chances of heavy late consequences conditioned on an economizing immediate decision.

(5) IF person is old
    THEN person is ill
    IF person is ill
    THEN person is under treatment

Putting apart the fuzziness in the antecedents above, these two rules exhibit another type of incertitude: they arc "generally true".

Two approaches can be envisaged:

– either the possibilistic one, based on conditional possibilities:

    Π(person is ill | person is old) = 1, and
    Π(person is not ill | person is old) ≥ 0

Π(person is under treatment | person is ill) = 1, and
Π(person is not under treatment | person is ill) ≥ 0
and on the Prade's computation;

– or the fuzzy one, using quantifiers.

As the first rule is concerned, people have the feeling that "most old persons are ill". Thus the first rule is "mostly true", i.e.

verit(rule 1) = 0.8 ("mostly").

For the second rule, people feel that "almost all ill persons are under treatment". Thus the second rule is "almost true", i.e.

verit(rule 2) = 0.9 ("almost").

The classical syllogism allows us to derive, in classical logic, the compound rule

IF person is old
THEN person is under treatment.

A very simple calculus

verit(compound rule) = verit(rule2) $\cdot$ verit(rule 1) = 0.9 $\cdot$ 0.8 = 0.72

(value interpreted as "frequently") allows us to appreciate that "frequently old persons are under treatment".

10.3 Perception-Based Theory

In medical practice (in most other decision contexts also) the information at hand is a mixture of measurements and perceptions. When all is measured, classical probability theory is of great help. However, it is not helping us to deal with perception-based information.

To be able to compute with perception-based information we first need a representation of the respective meaning in a form that is suitable to computing. After this representation is achieved, truth propagation from premises to conclusions can be put at work by a suitable logical engine.

Let us present in short the approach of [Zadeh 2002], the so-called CTP (Computational Theory of Perception). This approach involves a family of predicates (Zadeh calls them variable copulas)

isα

where α represents one of the following "symbols":

_ ("void"), for possibilistic constraints,
=, for equalities,
<, for inequalities,
p, for probabilistic constraints,
u, for usuality constraints,
v, for veristic constraints (expressed by certainty factors),
fg, for fuzzy graph constraints, etc.

The predicates (variable copulas) isα are used together with variables X and with constraining relations A to form the so-called **unconditional constraints**

X isα A

that are fundamental for "computing with perceptions".

(The variable X in the unconditional constraint above may have a structure, may be a function $f(Y)$ of another variable Y, may be conditioned by another variable, $X \mid Y$, may be multi-dimensional.)

Thus, X is D expresses the fact that the variable X has possibility distribution D, X is$p$ D expresses the fact that the variable X has probability distribution D, X is$u$ U expresses the fact that X is usually U, X isfg G expresses the fact that X is a function and G is its fuzzy graph, and so on.

Generalized constraints are the main semantic elements of the meaning-representation language. A **generalized constraint** is obtained from unconditional constraints as a production rule

IF X isα A THEN Y isβ B (equivalent form "Y isβ B WHEN X isα A")

or as an exception-qualified rule

X isα A UNLESS Y isβ B.

The Generalized Constraint Language is generated by combination, qualification and propagation of generalized constraints.

The rules that govern generalized constraint propagation in CTP coincide with the rules of inference in fuzzy logic. The basic rules, expressed in their most general form, are the following.

First conjunctive rule:	from	X isα A
	and	X isβ B
	we infer	X isγ C
Second conjunctive rule:	from	X isα A
	and	Y isβ B
	we infer	(X,Y) isγ C
Projective rule:	from	(X,Y) isα A
	we infer	Y isβ B
Surjective rule:	from	X isα A
	we infer	(X,Y) isβ B
Inversive rule:	from	$f(X)$ isα A
	we infer	X isβ B
	where $f(X)$ is a function of X.	

In the first conjunctive rule the copula isγ may be obtained from copulas isα and isβ by formulas analogous to that of operation * in Section 6.2.

From these basic rules other general rules may be derived, as exemplified by the following proposition.

Proposition. If the second conjunctive rule and the projective rule above are accepted, then we are entitled to use also the compositional rule:

$$
\begin{array}{ll}
\text{from} & X \text{ is}\alpha\ A \\
\text{and} & (X,Y)\ \text{is}\beta\ B \\
\hline
\text{we infer} & Y \text{ is}\varepsilon\ E.
\end{array}
$$

Proof. Indeed, let us suppose we know X isα A and (X,Y) isβ B. Then, using the second conjunctive rule, we infer

$$(X,(X,Y))\ \text{is}\gamma\ C.$$

Now, using the projective rule we infer

$$(X,Y)\ \text{is}\delta\ D$$

and using again the projective rule we finally infer

$$Y \text{ is}\varepsilon\ E.$$

This general rule is of little help. However, particular cases are very important. Let us give only two examples.

Example 1. A and B are fuzzy sets and $\alpha = \beta = \_$ (void). In this case B is in fact a fuzzy relation. We may consider $\varepsilon = \_$ and $E = A \circ B$, the composition defined in Section 9.1.

Example 2. A and B are probability distributions and $\alpha = \beta = p$. If X is uni-variate, A is 1-dimensional and if Y is also uni-variate, then B is 2-dimensional. We may consider $\varepsilon = p$ and $E = A \circ B$, the composition of distributions.

10.4 Solved Exercises

1) Express the following two production rules into a programmable form:

When a patient has significantly elevated troponin concentrations and an abnormal electrocardiogram, it is likely the patient has had a heart attack.

When a patient with chest pain and/or stable angina has normal troponin, creatin phosphaze and CK-MB concentrations, it is likely that the heart has not been injured.

2) [Zadeh 2002] We know:

> Most Swedes are tall,
> Most Swedes are blond.

How many Swedes are tall and blond?

Solutions

1) The two sentences are typical for medical knowledge. A lot of fuzziness and subjective feelings are expressed. A possibility to express such knowledge into programmable form is as follows:

IF   TC(x) is SE
       AND ECG(x) is NOT NORMAL,
THEN P(HS(x) is HEART ATTACK) is VERY HIGH.
IF   (CP(x) is CERTAIN OR SA(x) is CERTAIN)
       AND T(x) is NORMAL
       AND CK(x) is NORMAL
       AND CK-MB(x) is NORMAL.
THEN P(HS(x) is HEART ATTACK) is EXTREMELY LOW.

However, in these rules there are different fuzzy sets NORMAL. For CK concentrations NORMAL levels expressed in units/liter is a fuzzy set trapmf(0, 0, 0.16, 0.25). For troponin concentrations there is not yet a good NORMAL description!

2) A genuine expression of the sentence

Most Swedes are tall

is obtained considering "most" as a fuzzy set in the interval [0, 1]. Of course, the meaning of "most" is usually translated into a fuzzy set of the shape in Figure 10.3.

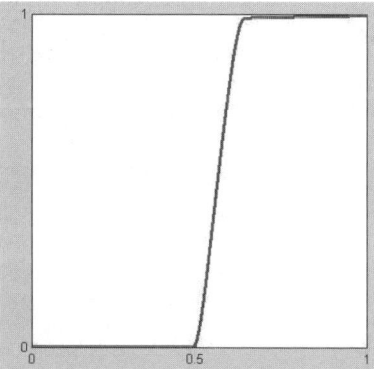

Fig. 10.3. Fuzzy set *most*

"Swedes are tall" is expressed as a conditional fuzzy set $T \mid S$ defined over the universe H of humans. (Indeed, for each human being h, a membership value $\mu_T(h)$ measures how tall h is, and another membership value $\mu_S(h)$ evaluates the Swede nationality.)

The sigma-count of the conditional fuzzy set is

$$\Sigma\#(T\mid S) = \frac{\sum_{h} \min(\mu_T(h), \mu_S(h))}{\sum_{h} \mu_S(h)}$$

and the translation of our sentence "Most Swedes are tall" is

$$\Sigma\#(T\mid S) \text{ is } most.$$

Similarly, "Most Swedes are blond" is translated into (where B is obviously the fuzzy set "blond")

$$\Sigma\#(B\mid S) \text{ is } most.$$

Taking into account the properties of the sigma-count (see Exercise IX.11), for the possible values ≥ 0.5 of *most*, the following inequalities are valid

$$2most - 1 \leq \Sigma\#(T \cap B\mid S) \leq most.$$

Psychological considerations advise us, when lacking specific information, to be cautious. We have no reason to favor either bound; instead, the median seems the favorite choice.

Thus, for a particular value ≥ 0.5 of *most*, after attaching a linguistic value to the median $most - \dfrac{1 - most}{2}$, let us say "frequently", we are entitled to express the sentence "frequently a Swede is tall and blond".

References

[1] Armitage, P., Berry, G.: Statistical Methods in Medical Research. Blackwell, Malden (1987)

[2] Bander, E.: Mathematical Methods in Artificial Intelligence. IEEE Computer Society Press, Los Alamitos (1996)

[3] van Bemmel, J., Musen, M.A.: Handbook of Medical Informatics. Springer, Heidelberg (1997)

[4] Bouchon-Meunier, B., Coletti, G., Marsala, C.: Independence and possibilistic conditioning. Annals of Math. and Artificial Intelligence 35, 107–124 (2002)

[5] Clancey, W.J., Shortlife, E.H. (eds.): Readings in Medical Artificial Intelligence. The First Decade. Reading Mass. Addison-Wesley, Reading (1984)

[6] Cramér, H.: The Elements of Probability Theory. Wiley, Chichester (1955)

[7] Daly, L.E., Bourke, G.J., McGilvray, J.: Interpretation and Uses of Medical Statistics, 4th edn. Blackwell Scientific Publ., Oxford (1991)

[8] Degoulet, P., Fieschi, M.: Introduction to Clinical Informatics. Springer, Heidelberg (1999)

[9] Dempster, A.P.: Upper and lower probabilities induced by multi-valued mapping. Annals of Math. Statistics 38, 325–339 (1967)

[10] Doll, R., Pygott, F.: Factors influencing the rate of healing of gastric ulcers admission to hospital, phenobarbitone, and ascorbic acid. Lancet 1, 171–175 (1952)

[11] Doob, J.L.: MeasureTheory. Springer, Heidelberg (1994)

[12] Dubois, D., Prade, H.: An introduction to possibilistic and fuzzy logics. In: Smets, P., Mamdani, E.H., Dubois, D., Prade, H. (eds.) Non-Standard Logics for Automated Reasoning, pp. 287–326. Academic Press, New York (1988)

[13] Dubois, D., Prade, H.: Fuzzy sets in approximate reasoning. I. Inferenced with possibility distributions. Fuzzy Sets and Systems 40, 143–202 (1991)

[14] Fu, K.-S., Shimura, M., Tanaka, K., Zadeh, L.A. (eds.): Fuzzy Sets and Their Applications to Cognitive and Decision Processes. Academic Press, London (1975)

[15] Giarratano, J., Riley, G.: Expert Systems. PWS Publishing Company (1998)

[16] Hakel, M.D.: How often is often? Amer. Psychologist 23, 533–534 (1968)

[17] Halmos, P.R.: MeasureTheory. Springer, Heidelberg (1974) (2nd printing)

[18] Hripcsak, G., Heitjan, D.F.: Measuring agreement in medical informatics reliability studies. J. Biomedical Informatics 35, 99–110 (2002)

[19] Jackson Jr, P.C.: Introduction to Artificial Intelligence. Mason/Charter Publ., New York (1974)

[20] Khoukhi, F.: Approche logico-symbolique dans le traitement des connaissances incertaines et imprécises dans les systèmes à base de connaissances. Thèse de doctorat, Univ. Reims Champagne-Ardenne (April 4, 1996)

[21] Klawoon, F., Smets, P.: The dynamic of belief in the transferable belief model and specialization-generalization matrices. In: Dubois, D., Wellman, M.P., d'Ambrose, B., Smets, P. (eds.) Uncertainty in AI 1992, pp. 130–137. Morgan Kaufmann, San Marco (1992)

[22] Klir, G.J., Yuan, B.: Fuzzy Sets and Fuzzy Logic. Prentice Hall, Englewood Cliffs (1995)

[23] Kolmogorov, A.N.: Grundbegriffe der Wahrscheinlichkeitsrechnung. Springer, Berlin (1933); English translation: Foundations of the Theory of Probability. Chelsea, New York (1956)

[24] König, H.: Measure and Integration. In: An Advanced Course in Basic Procedures and Applications. Springer, Berlin (1997)

[25] Kruse, R., Schwecke, E.: Specializations: a new concept for uncertainty handling with belief function. Int. J. Gen Systems 18, 49–60 (1990)

[26] Larson, H.J.: Introduction to the Theory of Statistics. Wiley, New York (1973)

[27] Lauritzen, S.L., Spiegelhalter, D.J.: Local computations with probabilities in graphical structures and their applications to expert systems. J. Royal Statistics Society Series B 50(2), 157–194 (1988)

[28] Lindley, D.V.: Introduction to Probability and Statistics From a Bayesian View-point. Part 1, Probability; Part 2, Inference. University Printing House, Cambridge (1965)

[29] Luger, G.F.: Artificial Intelligence. Addison-Wesley, Reading (2002)

[30] Mann, H.B., Whitney, D.R.: On a test on whether one of two random variables is stochastically larger than the other. Ann. Math. Stat. 18, 50–60 (1947)

[31] Minsky, M.: A framework for representing knowledge. In: Winston, P. (ed.) The Psychology of Computer Vision, pp. 211–277. McGraw Hill, New York (1975)

[32] Nakao, M.A., Axelrod, S.: Numbers are better than words: Verbal specifications of frequency have no place in medicine. Am. J. Med. 74, 1061–1065 (1983)

[33] Nilsson, N.J.: Artificial Intelligence. Morgan Kaufmann Publ., San Francisco (1998)

[34] Negnevitsky, M.: Artificial Intelligence. A Guide to Intelligent Systems. Addison-Wesley, London (2002)

[35] Pearson, K.: On the criterion that a given system of deviations from the probable in the case of a correlated system of variables is such that it can be reasonably supposed to have arisen from random sampling. Philosophical Magazine 50, 157–175 (1900)

[36] Poole, D., Mackworth, A., Goebel, R.: Computational Intelligence. Oxford University Press, Oxford (1998)

[37] Popper, K.: The Logic of Scientific Discovery. Basic Books, New York (1959)

[38] O'Reilly, R.C., Munakata, Y.: Computational Explorations in Cognitive Neuroscience. MIT Press, Cambridge (2000)

[39] Reghis, M., Roventa, E.: Classical and Fuzzy Concepts in Mathematical Logic and Applications. CRT Press (1998)

[40] Rich, E., Knight, K.: Artificial Intelligence. McGraw Hill, New York (1991)

[41] Rowe, N.C.: Artificial Intelligence Through Prolog. Prentice Hall, Englewood Cliffs (1988)

[42] Russel, S., Norwig, P.: Artificial Intelligence. Prentice Hall, Englewood Cliffs (2003)

[43] Savage, L.J.: The Foundation of Statistics. Wiley, New York (1954)

[44] Shafer, G.: A Mathematical Theory of Evidence. Princeton Univ. Press, Princeton (1976)

[45] Shortliffe, E.H.: Computer-Based Medical Consultations: MYCIN. Amer. Elsevier, New York (1976)

[46] Shortliffe, E.H., Buchanan, B.G.: A model of inexact reasoning in medicine. Math. Biosciences 23, 351–379 (1975)

[47] Shortliffe, E.H., Davis, R., Axline, S.G., Buchanan, B.G., Green, C.G., Cohen, S.N.: Computer-based consultations in clinical therapeutics. Explanation and rule acquisition capabilities of the MYCIN system. Computers and Bio-medical Research 8, 303–320 (1975)

[48] Simpson, R.: The specific meaning of certain terms indicating differing degrees of frequency. The Quarterly J. of Speech 30, 328–330 (1944)

[49] Smets, Ph.: Un modèle mathématico-statistique stimulant le processus du diagnostic médical. Ph. D. Thesis, Univ. Libre de Bruxelles (1978)

[50] Smets, Ph.: Belief functions. In: Smets, Ph., Mamdani, A., Dubois, D., Prade, H. (eds.) Non standard logics for automated reasoning, pp. 253–286. Academic Press, London (1988)

[51] Smets, Ph.: Belief functions: The disjunctive rule of combination and the generalized Bayesian theorem. Internet. Approx. Reason. 9, 1–35 (1993)

[52] Smets, Ph., Kennes, R.: The transferable belief model. Artificial Intelligence 66, 191–234 (1994)

[53] Smets, Ph.: The transferable belief model for quantified belief representations. In: Gabbay, D.M., Smets, Ph. (eds.) Handbook of Defeasible Reasoning and Uncertainty Management Systems, Quantified Representation of Uncertainty and Imprecision, vol. 1, pp. 267–301. Kluwer Academic Publishers, Dordrecht (1998)

[54] Smets, Ph.: Numerical representation of uncertainty. In: Gabbay, D.M., Smets, Ph. (eds.) Handbook of Defeasible Reasoning and Uncertainty Management Systems Belief Change, vol. 3, pp. 265–309. Kluwer Academic Publishers, Dordrecht (1998)

[55] Stoer, J., Bulirsch, R.: Introduction to Numerical Analysis. Springer, Heidelberg (1980)

[56] Taylor, J.C.: An Introduction to Measure and Probability. Springer, Heidelberg (1997)

[57] Trillas, E., Valverde, I.: On some functionally expressible implications for fuzzy set theory. In: Proc. 3rd Internat. Seminar on Fuzzy Set Theory, Linz, Austria, pp. 173–190 (1981)

[58] Voorbraak, F.: As far as I know: epistemic logic and uncertainty. Ph. D. Thesis, Utrecht Univ. (1993)

[59] Wang, Z., Klir, G.J.: Fuzzy Measure Theory. Plenum Press, New York (1992)

[60] Wilcoxon, F.: Individual comparisons by ranking methods. Biometrics 1, 80–83 (1945)

[61] Xu, H., Smets, P.: Reasoning in evidential networks with conditional belief functions. Int. J. Approx. Reasoning 14, 155–185 (1996)

[62] Zadeh, L.: Fuzzy sets. Inf. Control 8, 338–353 (1965)

[63] Zadeh, L.: Fuzzy sets as a basis for a theory of possibility. Fuzzy Sets and Systems 1, 3–22 (1978)

[64] Zadeh, L.: Toward a perception-based theory of probabilistic reasoning with imprecise probabilities. J. Statistical Planning Inference 105, 233–264 (2002)

Index

Printing: Krips bv, Meppel, The Netherlands
Binding: Stürtz, Würzburg, Germany